THE CLIMATE TRAP

The Climate Trap

A Perilous Tripping of Earth's Natural Freeze Protection System

Melvin J. Visser

Fortitude Graphic Design and Printing
Kalamazoo, Michigan

Copyright © 2010 by Melvin J. Visser
All rights reserved. No portion of this book may be reproduced, stored in a retrieval system, or transmitted in any form or by any means (electronic, mechanical, photocopy, recording, or any other) except for brief quotation in printed reviews, without the permission of the publisher.

Fortitude Graphic Design and Printing • Kalamazoo, Michigan 49001
www.fortitudegdp.com

Printed and bound in the United States of America

Visser, Melvin J. 1938- 2021

The Climate Trap: a perilous tripping of earth's natural freeze protection system / Melvin J. Visser

ISBN 1453627030
EAN 9781453627037

Includes scientific references and index.
1. Global warming. 2. Climate change. 3. Climate history

Graphic Design and Printing by Sean Hollins, Fortitude Graphics
www.fortitudegdp.com,

Cover design and interior illustrations by Chad Sutton, Bluerise Studio, www.bluerisestudio.com

Cover photograph of Coberg Island by Melvin J. Visser

Dedication

To the memory of Jacob Visser, (1898-1964) a wise and loving father. Of his many instructive sayings, I heard the following most often during this work. "Ideas are like leftover food in the ice box. If they do not smell right, they should not be swallowed."

Foreword

A mere five centuries ago, Magellan sailed around a world most people thought was flat to open trade routes providing great wealth and global power to the developed nations of his time. Gold and silver was plundered from heathens who were killed or enslaved and forced by the sword to serve the conqueror's god. At that time 500,000,000 humans lived on Earth and salt, silk, dyes, and spices were hot items of trade.

In recent decades, global governments have facilitated trade in a world whose god appears to be "economic progress." Goods and information flow unrestrained around the world in giant ships, planes, glass fibers, and microwaves. The developed nations now seek a "post industrial" lifestyle with developing nations supplying them food, hard goods, and toys at low prices. This system is performing beyond belief ... and it does not require enslavement or genocide ... as everybody wins. Once starving hundreds of millions join the global middle class and the current starving migrate to cities or immigrate to lands of opportunity. Now, 6,800,000,000 humans reside on the planet. Consumption of resources and energy, or more politically correct ... improving gross domestic products at an ever-increasing rate ... is a universal goal.

Is this sustainable? Can Earth support our human success? Will supplying energy for this highly successful human economic system change Earth's atmosphere and cook us off the planet?

Shortly after Magellan's time, Galileo used his powers of logic and a telescope to determine that Earth revolved around the sun ... a pursuit that resulted in his arrest and imprisonment for challenging existing religious beliefs. Adventurers and scientists continued to risk life and censorship to study Earth, the solar system, and the ever-expanding universe. Understanding of Earth and the cosmos has accelerated as space travel, satellite telescopes, and computers enhance our scientific capability. In less than a human lifetime, Earth's age has grown to 4.54

billion years old. We now realize that our Earth is a small speck in an ever-expanding universe of three times that age.

Currently, an assemblage of global scientists is convinced that man is influencing Earth's climate and will initiate intolerable warming within decades or generations. As throughout history, they are encountering religious and political objections. The global warming argument places us in the position of making a frightful choice for our progeny. Will we expend tremendous resources to change our energy sources on the chance that man is overheating the planet, or will we continue our current path and possibly leave them with an unsustainable and chaotic planet? This question certainly deserves our serious study and thought.

Examination of Earth's historical climate, scientific observations, and computer models are used by modern science to predict future climate. As seen on the back cover of this book, Earth's temperature has varied widely over the past 450,000 years. Earth has provided a friendly temperature allowing humans to propagate and prosper without migration for only for the past 12,000 years, a flash of time in its 4,540,000,000-year history. Scientists 'read' the temperature record for the past 450,000 years from an Antarctic ice core. Other scientists found that Earth was repeatedly glaciated very early in its history ... more than 2,000,000,000 years ago. Between its early freezings and the modern era freezings, Earth warmed to the point that Arctic areas were covered in tropical splendor. We can only imagine what the current ice-buried Antarctic continent was like at that time. Science has done an amazing job of determining WHAT Earth's temperature was WHEN, but the reasons WHY and HOW Earth's temperature varies are just beginning to be understood.

The primary forces, other than atmospheric composition, for determining Earth's temperature are its distance from the sun, the sun's energy output, and the reflectivity of Earth's surface. These items, and the wobble of Earth's spinning axis, vary in cycles from decades to tens of thousands of years. When the cycles align, Earth cools and polar ice caps grow. Ice reflects the sun's energy and cooler oceans adsorb more carbon dioxide from the atmosphere to further enhance cooling. When partially glaciated, the loss of energy due to Earth's reflectivity overwhelms the small energy input differences due to Earth's orbit and tilt. The start of glaciation should be a one-way trip to the freezer, but Earth suddenly and rapidly warms. WHY? The current belief is that the oceans warm and release carbon dioxide. WHY? HOW? What process would lead to a rapid heating? HOW could Earth freeze two million years ago? Wouldn't the atmosphere have been full of greenhouse gases at that time? There was no life on terrestrial Earth. Where was the carbon that eventually ended up in plants, animals and coal?

If we had a better understanding of the WHY and HOW of Earth's temperature history instead of just WHAT the temperature was WHEN, could we assess the future more clearly? These questions need answering. Earth does not magically change. Science should supply reasons that make common sense.

To address these questions, I spent a decade reading, consulting with experts, and piecing together "mind experiments" to seek logical reasons for a highly reflective shiny ball flying through space to suddenly heat up. By incorporating space age findings of the history of our universe and the latest oceanic science, I was finally able to go back to the early formation of our sustaining planet and follow the flow of carbon in and out of terrestrial, oceanic, and atmospheric compartments in a manner that made sense.

To aid my "mind experiments," I enlisted the fantasy help of an all-knowing genie. As my middle-of-a-sleepless-night thoughts led to yet another dead end, I would dialog with the imaginary genie. He grew to be a great help, and as time went I grew to seeing him as a spiritual companion … a protector of the planet. I tell the following story along with Ralph, the Earth Keeper, who takes me back in time for a front row seat to Earth's history.

Among many other interesting findings, we determined that Earth's bouncing out of an ice age is guaranteed. Earth has an amazing built in freeze protection system. But this system may backfire once the oceans start to warm … as it has before. Are we close to being caught in this irreversible pitfall … *The Climate Trap?*

Please fasten your seatbelt, open your mind, and join Ralph and me for a short trip through the billions of years of development of our splendorous home. It's a fast trip covering a lot of new territory, so I'll come back at the end to talk about what this history means to our uncertain future.

The Test Drive

Chapter 1
Cancun Mexico
March 2009

AFTER A DELIGHTFUL AND TIMELY flight on Mexicana Air, Gloria and I finally made it through the horrendous switchback of sweating and impatient humanity plodding through customs, gathered our luggage, and joined the long line exiting through the inspection station. Fortunately, we were experienced and knew that this short-term purgatory was well worth the heaven to follow. Saturday arrivals in Cancun have overwhelmed the airport for each of the twenty-five years we experienced it. The facility has grown and modernized, but not fast enough to keep up with the exploding Cancun and Mexican Rivera tourism.

For me, the inspection station is an especially anxiety-producing event. There is no profiling on entering Mexico. You push a button and a random red light means a total and thorough luggage inspection. We bring lots of luggage for our five-week stay and the anticipated hassle of watching it pawed through it is unnerving. I do the manly thing and let Gloria push the button. She got the green light.

An assertive and smiling porter beat his competitors to our side and threw our bags onto a dolly. He led us through a gauntlet of aggressive transportation, tour, and condo salesmen like the drum major of a small band. We snaked our way between the milling knots of sun seekers exiting in masses to a chaotic collection of giant tour busses, vans and taxies; all with their drivers shouting directions and competing for fares in hand-waving Spanish.

It felt good to be back. We shed our Michigan sweaters and took in a welcome bit of sun. The warm, humid Cancun afternoon heat was quickly replaced by the frigid air conditioning of our prearranged Thomas Moore van speeding to the Royal Resorts with eight eager vacationers.

We marveled at the beautiful palm lined and manicured boulevard leading to the beach. Our first trips were traversed on a bumpy lane with ragged paving. Then, Cancun's now famous seven-mile shallow crescent of brilliant coral sand beach contained a couple of hotels at the end now known as the "Convention Center," a rustic Club Med at the end we were coming to and the president of Mexico's thatched "summer home" in the middle. Good friends invited Gloria and I and our college sophomore daughter, Lori, to spend her spring break with them. Their condo developer was opening up development of the beach and took us for a bus ride to swim, play volleyball, eat chicken and view a mockup of "The Royal Mayan," a 200 unit horseshoe shaped project. We enjoyed Cancun's people, weather, beach, and food so much that we were hooked. We bought a penthouse unit at preconstruction prices and have been there every year since.

Now, except for the public park areas wisely preserved by the Mexican government, the beach is covered with billions of dollars worth of international hotels, condos, and luxury living. The area has bounced back from two major hurricanes and remained vibrant. By the time I had retired, our condo developers had completed five projects in the area and we had increased our holdings to five weeks.

Daniel welcomed us at the curb as we pulled up to the Royal Mayan lobby. "Welcome home amigos," was his warm, smiling greeting, and he meant it. Daniel had come out of a nearby Yucatan jungle village to scrub swimming pools when the Mayan opened and was now Bell Captain. Lucas, a former gardener, welcomed us at the front desk.

Federico, a tall and serious young man who took pride in his growing family and his job of taking care of the penthouse units was smiling at our door. "Buena suerte amigos, no fumar nada," is what I think he said. Through gestures and broken use of each other's language, we learned that the previous occupants, heavy smokers, didn't come this year. We could breathe fresh air and not have to open the unit for days to clear out the dead smell of stale tobacco smoke. Federico knew our dislike of the odor and always cleaned and aired our suite as soon as it was vacated.

When we first encountered the pleasant and serviceable Mayans of Mexico, we didn't think it could last. After nearly twenty-five years and going from dwellers of small villages to residents of a city of hundreds of thousands, they still enjoy providing friendly service with pride. There is nothing like it in other vacationlands. Mayan hospitality remains a large part of Cancun's charm.

It was getting late and we were starving, so we went down to poolside for the last rays of sun, happy hour, and a grouper sandwich. Gloria seldom eats fish at home, but can't wait for Cancun grouper. Joe, Jeff and Dale, long time Cancun friends who had arrived the week before, joined us for happy hour while their wives were getting ready for dinner.

Last year my discussions with these boys had gotten a little testy as we discussed global warming. I was glad to see they were still willing to talk. We exchanged pleasantries, got caught up on kids and grandkids, the current favorite eating spots, and last week's weather. When Jeff mentioned that it had been especially cool, Joe chimed in with this being proof that global warming was a hoax, and Dale asked if I was still swallowing everything Al Gore was offering. I saw Gloria's eyes roll as I opened my mouth in defense. Why can't friends be as welcoming as the staff? I thought. I didn't want to rekindle an argument at this time. We were saved by the arrival of their wives and our grouper.

"I'm glad you didn't stoke that argument again," Gloria said knowing how difficult it is for me to resist getting carried away. "Thank you."

"I'm sure we'll get into it later. I think that arguing in Cancun happens easily as it is much safer than arguing with family or coworkers at home."

"Leave me out of it." Gloria savored a large bite of grouper and smiled. "I have more enjoyable priorities."

The essence of their argument was a belief that Earth's temperature had varied widely during its history and temperature change was natural. Man was doing nothing to influence temperature and it was sheer arrogance to think that man could overwhelm nature. What would be would be, or in Mexico ... que sara, sara ... so relax and let tomorrow's tee time be your major concern.

My thoughts of global warming were rather simple. Greenhouse gasses were like a blanket and the more blanket you have the warmer you stayed. Man had increased greenhouse gasses and should expect a warmer Earth. The argument took on a deeper flavor when we discussed ice ages. To them, ice ages happened naturally and recovery from them was just as natural. From their conservative perspective, the Earth heated up and drove carbon dioxide out of the ocean to give the greenhouse gasses found in the ice cores. They identified it as liberal nonsense that greenhouse gasses initiated the thawing process.

I do not subscribe to a conservative or liberal ideology and I approached the problem scientifically ... but I didn't have the answer. To me, Earth's recovery from an ice age was a major scientific problem. When Spaceship Earth turned into a reflective white ball, there was no understandable source of greenhouse gasses or heat of the required magnitude to turn it around and melt ice. Yet Earth froze slowly and then bounced back to warm up rapidly. This could not happen with the minor changes in Earth's wobble or orbit. Earth heated quickly and the reason was not understood. How can we speculate on the future when there is such a large gap in the understanding of our past? Scientists have rather recently discovered that Earth was heavily glaciated two billion years ago, when the sun was much weaker. Where were the greenhouse gasses then and how did Earth recover from its freeze instead of ending up like Mars?

I bit into my grouper. It was delicious. Gloria and I spent the rest of the evening savoring the meal in peaceful quiet. I was retired and needed to enjoy a couple of weeks of relaxing before our daughter Lori and her three daughters quickened our pace. Last year Lori and the eleven and thirteen-year-old introduced me to zip lines and cliff rappelling. I had to get ready for whatever this year would bring.

I woke up before dawn and looked out at the ocean. A half a moon was tinting the dark ocean's wave tops with a lemon white lace. It was inviting. I put on a bathing suit, my favorite Tilley hat, and a little sunscreen. It was time for a five-mile round trip test walk to Club Med.

There are may ways to enjoy Cancun, but through the years, Lori and I have cherished our sunrise walks to Club Med. Cancun's seashore was, and remains, surprisingly natural with cruising schools of sardines and other small fish that attract larger fish, diving pelicans, and noisy flitting terns. Porpoises are frequently spotted out beyond the surf and high overhead the sky is patrolled by mysteriously shaped frigates.

In the early years, beachside construction of condos and hotels was a constant process. Multi-story wooden scaffolds covered with bareheaded, hard working Mayans in sandals would respond to a wave from Lori with appreciative whistles. Their construction equipment was crude, but their finished product beautiful and as lasting as their pyramids. The nature, culture, and construction progress walking tour became a cherished father/daughter experience during Lori's spring breaks from engineering school.

Lori continued to appreciate Cancun, honeymooned there two years after graduation, worked for six years, had her first child on her last day of work, and then returned frequently with her growing family. The pleasure of the ritual of our morning walk richened with time. Lori, like many teens of the time, stopped talking to me for what seemed like a long time, but made up for it when I matured. Our walks transitioned from discussion of college issues, to an interaction between practicing construction and environmental engineers, and then pleasant conversation between two parents. She stayed in shape and would be ready for the hike when she arrived. I told Gloria I wanted to check out the beach erosion, but I was just as worried about my stamina for five miles through wave and sand. I had to check out my systems and make sure I was up to the walk.

Hurricane Wilma had given Cancun a three day pounding four years earlier and removed much of the beach sand. Walking was impossible until sand was dredged back, and each year brought new challenges.

I left the condo, took deep breaths of ocean air, and enjoyed the glitzy view of Cancun's predawn sparkle from the ninth floor walkway to the beach elevator. On exiting to the ocean, I was glad to see the beach only a little thinned in front of the Royal Mayan. Our developer owned the two condos to the south and by the time I got to the last property, the Royal Islander, the beach was waning and in places waves lapped right up to the break wall. I picked my way through the rocks and for some reason my thoughts began to dwell on last year's arguments with Joe, Jeff, and Dale. I was wasting time stewing about the past. At a thin strip of remaining sand, I stopped, stretched and looked out over the ocean. Nature's stage was being set for what should be a glorious tropical sunrise. I had to get out of my stew and appreciate the present.

Just past the Royal Resorts properties, at an empty spot of beach used as a staging area for pipe and equipment during the dredging operation, I climbed the dune to view the sunrise in peace and quiet. There was still some pipe and debris scattered about. I looked all around and there was nobody in sight. This place was mine and mine alone for the next half hour and I would cherish it. The walk could wait. I leaned against the supports of an elevated fuel tank and scanned the horizon for porpoise.

There were no clouds at the horizon. The sun rose right out of the water to light the wave tips in fiery shades of orange and the clouds above in pinks and magentas. I stood watching the sunrise unfold and the various birds making their way to breakfast.

Suddenly I felt the uncomfortable presence of someone intruding on my space. Right beside me, leaning against one of the other support pipes, was a young man who had materialized out of nowhere. The sudden presence of this stranger did not arouse fear; in fact, a strange peace came over me.

"Beautiful, isn't it?" His voice was young, but calm, confident, and nonthreatening.

"Yes it is very beautiful," I answered, wondering where he had come from and what he was up to.

"I'm glad that you came by this morning." He sounded as though he was waiting for me to stop by. He was a slightly built man in his early twenties, of medium height, wiry, and rather handsome with a confident smile and clear light brown eyes. He had the typical Mexican uniform of black pants and white shirt. I could imagine him as a silver peddler that walked the beach all day, or perhaps one of the itinerant spiritualists that sometimes led groups on the beach, or maybe a condo salesman.

He laughed, showing a perfect set of shiny white teeth. "Yeah, I've been all of those and more."

Was he reading my mind?

"You seemed to have a little problem down there." He nodded toward the Royal Islander break wall.

I didn't have any problem that he could see from here. I may be triple his age, but I was handling myself.

"No, you would have no problem going to Club Med and back, I'm talking about your other problem."

"What problem?"

"Your dilemma about Earth's history and how it recovers from an ice age."

Wow, how did he know about that?

He smiled again. "I heard you arguing last year and have followed your quest. You really want to go back there to see how it happens?"

"I'd love to," I said and moved toward him. What kind of scam is this, I suddenly wondered. Condo sales? I turned to leave this nut.

He raised his hands palms out. "Honestly, no scam. I can take you back as far as you want to go."

"Who or what are you?" I wanted to leave, but was beginning to believe this whatever.

"Call me the Earth Keeper."

I scratched my head wondering what I was getting into. Was this foolishness or a chance of a lifetime? I decided to call his bluff.

Chapter 2

"Ready?" the Earth Keeper smiled his charming and confident smile.

Ready for more than you can possibly deliver, I thought. He had to be a phony, but how did he read my mind and know so much about me? How did he look so common that he could fit in anywhere, yet possess an uncommon ability to sense everything going on, even my thoughts. "I want to go way back."

"How far?"

"To when the Earth was young."

"The ice ages, right?"

If I had one chance, I really wanted to understand. "Further. When it was still hot ... too hot for oceans."

"Okay."

That's it? Just an okay. Is it that easy? I might as well get this over. "I want to go right now."

"You're not ready to go that far, and there is not enough time right now." He looked at the rising sun and the daily emergence of shell picking beach walkers. "We both have to get ready."

"I thought so. You're no Earth Keeper, or whatever, and I'm a sucker for talking to you." I turned to leave.

"Oh, you're ready to go back a few tens-of-thousands of years." He raised his right hand toward me and twitched his head back with a "give me five" smile.

As our hands touched, his eyes shone brightly and a flash of light blinded me. Did that light come from his eyes? When I could see again, I was in what appeared to be a cramped space ship sitting where the rusty elevated oil tank stood seconds ago. I was alone and surrounded by instruments with a half a dozen viewing ports around the perimeter. I was harnessed into a reclined bucket seat.

"Comfortable?" He sounded like he was right with me, but there was no room.

Now I believed him and a new set of fears surfaced. "I swallowed hard to get my heart out of my throat. How long will I be gone?" Gloria would be wondering where I was in a couple of hours.

"You can easily do all the ice ages in an hour. I'd suggest that you just do one for a test drive. When you come back, I've got a reading assignment for you. We'll do the big trip tonight. By the way, my friends call me Ralph."

I felt a slight sense of motion, then nothing. Ralph? What kind of Mexican name was that? I probably better start thinking of him as an alien, or extra world figure.

"Just think of me as Ralph, your guide," he laughed. "I've got you at 50,000 feet and holding." Ralph's comforting voice came through an overhead speaker or something that made him sound like he was right with me. "From here on, you're doing the driving."

"How?" I felt my voice quiver as lights blinked and what sounded like control motors hummed. The ship seemed to be begging to be used. "I have no idea what these controls are for."

"Just think about it and I'll answer."

I wondered how I controlled the time and as soon as I wondered, the left hand side of the control panel lit up and several readouts appeared under the overall heading of "TIME." "I see the indicators, now what do I do?"

"Set the one on the lower left to the time you want to go to. You better do it quickly, because I can't keep you out of sight for long. It's nearly daylight."

Now I felt like an idiot. I couldn't get my mind to focus on when the last ice age started. Was it millions or thousands of years ago?

"Just turn the lower left knob to the time you want to go to." Ralph was beginning to sound impatient.

I rotated the knob to the left and watched the digital readout clock back to minus 16,000 years.

"Good choice, you can watch it turn around from there." Ralph sounded relieved that I'd finally made a move. "Now think about how to get there."

I thought about it and wished I could get there quickly, and then come back slowly through time. The right hand side of the control panel lit up with several options under "TRAVEL."

"Press the button that says "Go To," It's on the upper right. After you do, look out the window and hold on."

I tentatively reached toward the button. What was going to happen? What did I have to gain on this crazy venture? What did I have to lose? I thought about how many times I had wished for the capability to see what really happened when the Earth was forming. I had everything to gain and nothing to lose. I punched the button.

The craft shook and bounced while its power station hummed and whined, seeming to cycle in and out of overload. The pungent odor of ozone tingled my nostrils and the cabin temperature raised a few degrees. The discomforts were minor as I watched history unfold beneath me. The modern structures of Cancun were wiped out in seconds. The Mayan culture rose and fell ... or in this direction, did it fall and rise? In what seemed to be a minute, there was nothing but jungle and ocean. Then the ocean started to disappear, the shoreline rapidly receded, all the way to where the deep waters of the Gulf Stream passed the Yucatan Peninsula.

"Are you all right? You're not saying anything." Ralph sounded comfortable now, probably happy that I was finally out of modern Cancun's sight.

"This is fantastic. But what's happening to the ocean? It's dried up clear out to the Gulf Stream."

"What time is it?"

I looked at the indicator on the right hand side. It was at 15,872 years and counting. "We're closing in on minus 16,000 years."

"What do you think Canada looks like now?"

"It should be covered with ice."

"And where does ice come from?"

"Water." I remembered that there were ancient artifacts found in caves far below the modern sea level. "The ocean levels went down hundreds of feet to supply the two mile thick glacier. I'm looking at the result ... I cannot believe it!"

"Well you are there. How was the ride?"

"A little rough and some of the systems seemed to be straining at times, but everything seems okay now." I peered across the expanse of beach to the new shoreline miles away.

"These older models are a little under powered and rough under stress, but plenty adequate for this short trip. Do you still want to see some ice?"

"Yes," I replied and as I thought of it a group of instruments under "LOCATION" illuminated in the middle of the console. The central display was a graphic of Earth, centered on the Yucatan. I did not see any way to select any other displays. Ralph read my mind, or whatever he did to know what I was thinking.

"Zoom out on the central graphic, and then move it in whatever direction you like. When you've decided, put the cursor on the location and hit the "Go To" button."

I set the cursor on what I thought would be south of Chicago and pushed the "Go To" button. The craft gained altitude and headed north. In the few seconds that it took to cross the Gulf of Mexico, I could see it was an emaciated puddle compared to the gulf that Mexicana Air had flown me over. From the southern U.S. coast inland, the landscape was covered with grasses amongst dead trees. Further north, the forest cover was intact.
"Enjoy the ride?" Ralph was still with me.

"Yes, but why didn't I feel the acceleration and breaking?"

"It's amazing what happens in controlled gravity. Are you where you want to be?"

I looked to the north and saw the great expanse of white. Its leading edge was thirty or so miles away. I settled into 50,000 feet and hovered. What a spectacular sight. The whole area I had grown up in was under a carpet of white as far as I could see. The carpet was thick … I'd heard estimates of two miles thick and it was believable.

There were piles of dirty ice at the base of a reasonably vertical cliff of variegated blue shades undulating to the north and south as it stretched to the eastern and western horizons. Far on the northern horizon, dark clouds with grey streaked bottoms were dumping snow. The glacier must be still advancing, I thought. "It would sure be nice to sit here and watch time advance by about a hundred years a minute," I said to myself.
"Just dial it in and punch the button." The "Rate of Time Travel" display illuminated and I started to follow Ralph's instructions. "I'd recommend 200 years a minute, we haven't got much time and you went back quite a way."

I tentatively punched the button and glued my eyes to the north.
The static glacier bounced forward like a monstrous unstoppable bulldozer in an old time movie. The cycles between day and night flickered at an imperceptible 1200 days per second, making the scene appear in eerie twilight with frequent lashings of storm clouds and snow filled white-outs. In general, the wall moved at a rather uniform speed. Blocks of ice broke

from the upper surface and joined the tumbling of dirt, trees, and ice chunks the size of skyscrapers that were pushed in front of, and eventually ground up by, the advancing sheet. A miles-long section would speed forward in a mass of fractured pieces and the rest of the wall would advance to assimilate the renegade pile. Then it suddenly changed.

Snowstorms became less frequent, the advance rate slowed, and the sheer front crumbled. Water started accumulating in the rubble and streams flowed out from under the sheet as far as the eye could see. I punched the stop button.

This was about as far as it is going to go. I wanted to stay in real time and go up for a better view. The "LOCATION" area of the control panel was displayed and an "ALTITUDE" dial illuminated. "I just dial in the altitude, right?"

"You're catching on, go up for a look."

I dialed in an altitude of twenty-five miles, then a location near Hudson Bay. After punching the "EXPRESS: GO TO" button, I was there in a heartbeat. There was no sensation of motion. From this altitude I could see nothing but white. The sun, low in the eastern sky, reflected off of hundreds of miles of melting ice. It was blinding. To the northwest, the ice field was still covered in the darkness of night. Even though I'm 12,000 years into the past, I'm still on Cancun time. I'd like to see this at high noon. Immediately, a dial labeled "ADJUST LOCAL TIME" was illuminated. I cranked the knob to move it from 07:15 AM to noon and my world brightened.

If we were on Cancun's hour, we must also be on the same month, I thought. Today was the spring equinox, so the sun should be lighting up the world from pole to pole. I could use the edge of dawn as a reference, get a good idea of where the North Pole was, and maybe go down for a closer look. I wondered how thick the ice was there. I asked for a three hundred mile high overview and was there in a second, enjoying an astronaut's view of Earth.

I looked south and located the expected green edge, but was surprised to see green to the west. I shouldn't have been, it was well known that Alaska was not glaciated in the last ice age.

I turned to the north, following the edge of darkness in the distance to get a focus on the North Pole. For hundreds of miles to the south of this edge, I was looking at dark open water. Far to the west a sooty fire rose out of the ocean like a belching volcano. "What?" I shouted to no one, and then I rubbed my eyes in disbelief at what I thought I saw.

"Anything wrong?" The ever-present Ralph was back.

"The Arctic Ocean is open." I blinked to look again ... "And it's on fire."

"And you expected?"

"Thick ice. Aren't we just coming out of an ice age?"

"Yes."

"Then how come the North Pole is melted already?"

"This will work better if you let me ask the questions."

"Okay, ask." I was beginning to trust this whatever.

"We just went through this above Cancun, but let's do it again. Where did all that ice down there come from?"

"Snow."

"Where did the snow come from?"

"Clouds."

"Where was the snow before it was in the clouds?"

"In water vapor in the air."

"And before that?"

"In the ocean ...there is no way all that ice could have formed without an open Arctic Ocean, or another nearby ocean with winds blowing towards shore."

"You've solved this mystery again, now what?"

Now what, I thought. Does this guy see oceans burning every day? "Ah ..." I stammered gesturing with my eyes toward the flaming ocean with billowing smoke covering the horizon. "There is a strange fire out there."

"Strange? It is much less strange than many things you will see tonight, so please do not dwell upon it on this practice run." The cabin filled with an icy silence and I felt that it would be useless to inquire further. Burning oceans not unusual? What was I in for tonight?

He broke the silence with a command "I'd suggest that you use this test drive to develop your piloting skills. You can use the same controls that got you here, or switch to joy stick."

A stick with a handle grip came out of the counsel. I grabbed it tentatively, was it like an airplane's?

"It's straight forward, but a little tricky. You push it horizontally in whatever direction you want to go, pull it up to go up, and push it down to go down. Your thumb has access to PAUSE and MAINTAIN keys. PAUSE will stop you in your tracks. MAINTAIN gives you options to maintain time, altitude, speed, or direction. It's all fail safe, so you won't crash."

I clicked my thumb on the maintain button and froze the time for a second, then clicked back to real time. "Should I freeze the speed?" I asked, thinking that I didn't want to go too fast on my initial solo flight.

"No need. In this model, speed is limited by altitude, so the lower you are the slower you will be able to go."

I moved the stick forward and pushed it down, heading toward the ice surface. At ten thousand feet, it slowed considerably, then at a thousand even more. I moved north and followed the northern edge of the open ocean. I watched the waves fly by and guessed that this pokey speed was about a thousand miles an hour. My thoughts were immediately confirmed by an illuminated speed indicator reading 1152 MPH. I went over the ice, dropped to fifty feet and thumbed the PAUSE button to hover over a small pool. Rivulets of water were spilling into the pool and it in turn was overflowing into a crack. The melting seemed fast. The outside temperature registered 85 degrees. The snow surface was blackened with soot. Wow, a warmer Arctic than I ever visited and dirty snow to reduce the reflectivity. What is the composition of the atmosphere I thought, hoping to bring up answers.

A display illuminated on the far right of the counsel. "ATMOSPHERIC COMPOSITION" was the group title, but the dozen or more readouts under it were displaying "Inactivated." Too bad, this could be interesting.

"Don't worry. I'll have a full compliment of analytics available tonight."

I had forgotten that this was just a test drive. I regained altitude and crisscrossed the icy expanse at high speed, then hovered over the Great Lakes area to watch their early birth and the raging melt waters exiting them. Gaining comfort with my new environment, I went low and cruised the perimeter of the receding glacier. On the southwestern edge, I lingered and finally saw a herd of mastodon drinking from the melt water. While at lower altitudes, my path was at times erratic, even when I was not influencing the controls. "What are the speed bumps I'm running into?"

"Humans. They get excitable when they see us, so we try to avoid them."

I went back to the northeast and watched the wet country of the Canadian Arctic come back to life. Within decades plants were reestablished and animals followed. I hurried over to watch the Bering Strait flood over and went up to the northeastern Canadian Arctic to watch the once continent shrouding glacier shrink to a pile of ice a few hundred miles square and a few hundred feet high. I was ready to return.

I called up the global graphic, confidently put the cursor on the tip of Isla Mujares, the most prominent recognizable feature near Cancun and pressed the "Go To" button. I was getting comfortable with the craft and my skill level. I looked forward to enjoying the ride while I thought about tonight's venture of unbelievable magnitude.

Over the Gulf of Mexico, Ralph's voice came on the speaker. "I see that you're coming home. Did you enjoy the trip?"

"Fantastic. I cannot wait for the big one."

"While you were watching the ice melt, I was getting ... WHERE are you going?" His voice level vibrated the cabin. Earth Keeper guides should remain calm. Was I crashing?

"I set the cursor on Isla Mujares, is that close enough?"

"You should have pressed the "HOME" button. Press it now ... no don't, I'll take over."

"What's wrong?"

"I was going to bring you in through a cover of low clouds. Isla Mujares is clear and there are a thousand kids out there celebrating the equinox at a Mayan shrine. You'll be seen."

The ship went flying by Isla Mujares then out to sea, down to sea level, and stopped. "I'm keeping you here until I get cloud cover. I'll create another diversion to bring you back."

After a few minutes, the craft went straight up through the low clouds, over to the beach and straight down. The top opened. "Hurry." Ralph frantically beckoned to me.

I climbed down a ladder that came out of nowhere and the spacecraft transformed back into the rusty abandoned tank.

"You made that pretty exciting. Sorry I didn't give you specific instructions for the return."

"Why is everyone running down the beach?"

"They're going to see groups of dolphins to the north and south of us."

"So nobody is looking up. Was I seen at Isla Mujares?"

"I'm sure, but we can handle that."

"How about the security guard heading this way?"

"He's the corporate guy watching this junk. Does your wife like silver?"

"She's nuts about the elephant bracelets the beach vendors peddle."

"You better buy her one," he said taking a bunch of bracelets from a wooden box and polishing them with a white cloth. He held one out, one Gloria would really like. My eyes brightened. Where was he going with this? I didn't have a peso in my pocket. "It's yours for eighty pesos," he said, winking at me. "You drive a hard bargain amigo."

I put my hand into my pocket and was going to show him it was empty... There was a wad of bills in it, a couple of hundreds and half dozen twenties."

"Everything all right here?" the guard addressed me in English while glaring at Ralph.

"Just fine officer." I paid Ralph.

"From now on, keep your business on the beach," he said in Spanish. "This is private property."

When he was out of sight I thought I'd test Ralph's sense of humor. I offered the bracelet back to Ralph. "You are wasting your time as a space travel guide, you could clean up selling on the beach."

Ralph laughed, "I see that you are impressed with the ingenious beach hawkers, so am I. Keep it, make Gloria happy."

I was very impressed with these outgoing and patient salesmen and considered them a part of Cancun's charm. Gloria deemed them essential. "What's the reading assignment you mentioned?"

"Go to the Royal Caribbean library. You'll see a Clive Cussler novel. Read it well enough to get the gist of the science presented."

"I'm going to learn science from an underwater novelist and somehow escape for the night to go billions of years back into history?" This was beginning to sound preposterous.

"No te preocupes amigo, trust me."

How did he know that my limited Spanish vocabulary included "don't worry?" But then how did all these strange things happen around him? I walked toward the Carib muttering "no te preocupes."

Chapter 3

THE ROYAL CARIBBEAN LIBRARY, SUCH as it is, was in the activity center. I waited for a break in a ping pong game and squeezed past a table full of kids painting tee shirts to access a couple of shelves of well-used books dropped off by guests to share with other guests. The picking was usually a motley selection of romance and action. Luckily, I spotted a Clive Cussler novel at the end of the second shelf. I assured myself it was the only one and pocketed it.

I thumbed through the book while walking back to our room. I'm a former SCUBA diver and have enjoyed several Cussler novels, but had not heard of *Fire Ice*. I am also quite sensitive to energy supplies and have a pretty good understanding of chemicals. Learning that fire ice was a common name for methane hydrate, a solid compound of methane and water with potential to fuel the economy for centuries turned me off. How could someone invent a chemical compound for a story? I wasn't eager to get into this reading assignment.

I had forgotten my key, so I stopped at the Royal Mayan desk. Luckily, Lucas was at the desk and gave me a key without even asking my room number. I was returning from my morning walk later than normal. Our room door opened just as I left the elevator and Gloria exited, dressed for the swimming pool.

She gave me a quizzical smile. "Where you been?"

"The usual, just a walk down to Club Med."

"Anything interesting?"

Only a trip to the last ice age I thought, then remembered what she would be really interested in. "I did find an interesting silver salesman." I dangled the elephant bracelet in front of her and she shrieked like a pre-teen.

"It's beautiful, what did you pay for it?"

"Eighty pesos," I said with pride, knowing it was a bargain.

"Eight bucks! What a deal. How many did you get?"

"He only had one."

"What was his name? What did he look like? I'll wait for him and order some more."

"He was selling out, leaving town, I think," I lied.

Gloria went back inside to stash her prize. She knew the bracelets were junk, but loved bargaining for, wearing, and giving them as presents. She reappeared with an inviting smile. "Want to join me for pool aerobics?"

"No, I'm a little tired from the walk. I'll stay at the poolside and read a novel I picked up at the Carib. I'll see you down there."

I found a spot in the shade to thumb through *Fire Ice*. It wasn't the best Cussler novel I'd read, but the methane hydrate was beginning to sound as real as Ralph's space ship. According to Cussler, when methane and water came into contact at low temperatures and high pressures, they combine to form a white solid. This was a real problem in Alaskan natural gas pipelines. The gas had to be desiccated to desert dryness to avoid the formation of filter and compressor clogging methane hydrate.

When methane formed from decaying vegetable matter on the ocean floor, or seeped up from gas and oil layers far under the ocean, methane hydrate formed. In Cussler's novel, the bad guys were going to disrupt deposits off the coast of New England to cause a tsunami.

Gloria dried off beside me after her aerobics were done. "How about a seafood pizza for lunch?"

"Good idea. I'll check email while you shop." This was one of Gloria's favorite ways to spend the mid day … browsing at the Mall after a crispy seafood pizza and a Corona at the La Fisheria. We had accumulated a house full of treasures found there.

After lunch, I hurried to the public Internet café. I suspected that Ralph wanted me to read *Fire Ice* because of the methane hydrate and I had to know why. I Googled "methane hydrate ocean" and got 157,251 results from a 0.17 second search. Methane hydrate was everywhere. It was first recognized from a well drilled in Siberia in 1964. Off the east coast of the U.S., an area the size of Rhode Island contained enough methane hydrate to supply natural gas needs for 70 years. The Department of Energy was calling it the energy source of the future. It was estimated that there was more carbon stored as methane hydrate than there was in all of Earth's coal and fossil fuels combined.

This was sure a humbling experience. I consider myself to be up to date on energy sources and the biggest carbon source in the world was unknown to me. I had no idea of the magnitude of underwater deposits of a solidified gas. Thanks Clive.

I studied a screen that displayed known deposits. Methane hydrate was found from the Arctic to the Antarctic and all latitudes in between. It was off shore of every continent. I stared at the graphic and thought there was no reason for it to be excluded from uncharted waters. It was likely that researchers had not scratched methane hydrate's surface.

I felt someone looking over my shoulder and through my ex-hockey referee peripheral vision identified a middle-aged male in shorts, sailing shirt and straw hat; another husband wandering around while his wife shopped.

"Amazing, isn't it?" The stranger sounded familiar.

"Sure is, this stuff is all over and I didn't know it existed."

"Seen the phase diagram yet?"

A phase diagram is a graph of temperature vs. pressure that identifies conditions at which a substance exists as a solid, liquid or a gas.

"That would be interesting. I know it is stable only under pressure, but I haven't seen much on temperature." I Googled "methane hydrate phase diagram."

"You'll be ready when you understand the phase diagram. I'll see you at six."

I turned around and extended my hand. "Ralph ..." I started to say, but there was nobody there. Across the room, Gloria stood at the front of a shop, facing me and wondering what I was gesturing. I pointed to my watch and raised five fingers. She raised five and pointed to the leather shop.

From the phase diagram, I found that methane hydrate had to be under at least 600 feet of freezing water and if the temperature got to only 50 degrees Fahrenheit, 2000 feet of water pressure would be required to keep it contained.

I was ready and eager to get going. I wondered how Ralph would get Gloria out of the way by six o'clock, but that was probably one of his minor concerns.

Chapter 4

ON THE RIDE BACK, THE sky was cloudless and the bus packed with spring breakers heading for the Oasis … a half a mile before our place. We planned to go down to the beach to read, but just as we entered the Royal Mayan people were flocking in to get out of the rain.

Gloria stepped back from the dripping crowd. "Where did that come from? It was beautiful all the way here."

"Those little storms can blow in from the ocean in minutes." I tried to sound knowledgeable, but I really wondered why Ralph called in the cover and what he was up to now.

"Did you bring this?" A dripping wet Jeff appeared beside us.

"It's probably got something to do with that UFO that buzzed Isla Mujeres this morning," Joe laughed at his side. "I've never seen a storm come out of nowhere like that."

"UFO?" Gloria questioned.

"Yeah, it buzzed a bunch of school kids by the shrine at the end of the island. The local TV is full of it."

Jeff planted a friendly elbow in my ribs. "You never know what will happen now that we've heated up the planet."

"I hope you guys don't rerun that argument," Gloria said. "It's clearing. I'm going to get some sun."

I followed her up to the room and out to her favorite reading perch overlooking the beach. I lack the ability to sit in the sun, so I excused myself after a few minutes. I'd have loved to see what the pipe yard looked like, but thought I better wait until six. I went to the beach and headed north.

On returning from a short walk, I saw that Gloria was still on her chair and took a long lazy swim. I just floated on my back and let the waves bounce me around, wondering what the evening's trip would bring. I was convinced that Ralph would produce and was eager to get on with it. Actually, I'd been obsessed with wonder about Earth's early atmosphere and carbon cycle for years. My career was spent in manufacturing chemicals, and then cleaning them from dumpsites and groundwater after the U.S. got sensitive to chemicals in the environment.

In my later career, I worked with governmental, academic, and consulting geologists as we developed methodologies for removing pollutants from the ground and ground water. They were amazed at my lack of knowledge of geology, and I equally amazed at their lack of knowledge of chemistry and biology. Somehow, my twisted mind started wondering what the atmosphere of the early Earth was like. I asked questions and did not get acceptable answers. Where did all the carbon come from that was eventually deposited as limestone (calcium carbonate,) coal, oil, and natural gas? Was it in the atmosphere at one time?

A couple of geologists helped by referring me to decades old "Definitive Papers" on the subject. These resources from the 1950s had the early atmosphere as very similar to our modern atmosphere. The earth cooled down, crusted over, and then volcanoes spewed forth the oceans and an atmosphere. I could not imagine a molten hot Earth crusting over and holding the oceans and atmosphere inside ... then leaking them out as Earth cooled. Scientists had certainly proven that our universe and Earth were old, but there was very little understanding of the pre-rock Earth. I felt that unless we had a better understanding of carbon's historical role in Earth processes, we would never be able to assess its role in the future.

In the past, I may have been more than a little arrogant as I used my slivers of chemical and engineering understanding to pick holes in the logic of geologists and astrophysicists as they slighted my interests. Now that Ralph and Clive had introduced me to methane hydrate, I was appropriately humbled. There may be a lot that the experts did not know, but in the large universe of knowledge needed to understand Earth's formation, I had barely scratched the surface. I hoped I could get to the right places and ask the right questions on tonight's trip. There would never be a better opportunity.

I saw Gloria leaving her chair and met her on the stairway to the elevator. "Got time for happy hour?" I asked, even though I wasn't about to drink anything before the ride of my life.

"I'm tired, whipped. I hope it isn't that old sun poisoning acting up again?"

I knew that Ralph, and not either of us, was under control of our fate. It was approaching 5:30.

At the room, Gloria decided to lie down and rest for a moment. Within five minutes, she was out like a light. I thumbed through a folder full of information I had collected on Earth's formation, and then put on a comfortable pair of shorts and a neutral shirt. I spotted a disposable camera on the dresser; saw that it had one shot left in it, and pocketed it. I just couldn't pass up the next opportunity I had for proof of a burning ocean to show Joe and Jeff and other skeptics. I headed down via the least likely to be seen route. It had cooled off and the beach was deserted. I walked the water line past the area where I'd meet Ralph and hooked back.

The area looked different. The old fuel tank was just as rusty, but it seemed much larger. I looked around and wondered if something had gone wrong.

"A beauty, Isn't she?" Ralph appeared behind me. I'd just scanned the area and there was nobody in sight.

"It's bigger. Is this a different ship?"

"With the Isla Mujeres show this morning, I called in a later model. It's easier to keep invisible."

"Sorry about that." I really felt bad for that mistake. "Will we be all right tonight?"

"We're all set. I talked the boss into a model with intergalactic capability. We will be restricted to Earth, but with what you want to see, we would have been straining the temperature limit, analytics, speed, and maneuvering capability of older models."

"We? Will you be with me this time?"

"Wouldn't miss it. I've only been there a couple of times."

What a break. This would be much better than being alone a billion years from home. I wondered what kind of operational support Ralph possessed. "Do you have a fleet of these ships?"

"I have whatever is needed to see what you want to see."

From the flat tone of his voice, I realized I was not about to know who or what was behind him. I'd better just stick with business. "Does it control like the last one?"

"Exactly. Now before we go, if you are piloting and want to go home, what are you going to do?"

"No te preocupes." I smiled. "I'll punch the RETURN HOME button. No more buzzing Isla Mujeres."

"You're ready, give me ten." He smiled and put his palms forward.

Early Formation

Chapter 5

WHEN I SLAPPED HIS HANDS, the flash of light from his eyes was less frightening, even welcoming this time. When my senses returned, the shock was greater than my previous awakening in a cramped craft with a few viewing ports. This ship was huge … and plush. It was like going from a 1950s VW Bug to a modern luxury vehicle. Ralph was at my right side, both of us in comfortable chairs that reclined and rotated to allow viewing of digital displays, electronic screens, and a multitude of ports around the craft's perimeter and overhead. There was plenty of room to walk around and a comfortable looking couch behind us. The air was fresh. Ralph was busy with what appeared to be a preflight procedure that seemed to be getting the desired responses from dozens of digital readouts and blinking lights.

"Appreciate the upgrade?" Ralph looked up and gave a smile that hinted of pride in his efforts. I viewed his capabilities as infinite, god-like, but he seemed humanly pleased with our space chariot. He must have spent the day pulling all the strings he could to supply my little desires with the best of his service. This was unbelievable.

"It's great! I appreciate the room, but where are the seat belts?"

"You won't need them on this trip. No humans to dodge, and this baby is smooth."

"Driven only by a little old lady on her way to church I suppose."

"And she was an atheist." After the joke, his tone changed. "Seriously, this venture will be at the limit of our capability. We are going to have to work together. I'll provide you the trip and answers to questions, but you will be in control. What you learn will depend on how hard you work. I am not capable of putting answers into your head."

I had only asked to be taken back for a first hand experience. "I understand."

"Ready?"

"Yes." I whistled out a lying lungful of pent up air. How did I know if I was ready?

"Remember how once you got to the ice age you wanted to know the conditions that preceded it?"

"Yes, I remember. I wanted to know the status of the atmosphere before the ice started melting." After seeing the fire in the polar ocean, there was a lot more I wanted to know.

"I'll take you a little further back than the hot Earth with no oceans you requested. You get one shot and I want to deliver satisfaction."

"Thanks. If we have the time, that's a great plan."

"I'll take you back express this time. I don't think it's worthwhile for you to see history run backwards."

"I'll gladly start at the beginning." I realized that I had stopped breathing and tenuously inhaled a long, slow breath.

"You'll black out on the express trip, so get comfortable," he nodded his head toward the couch. "I'll wake you up in a few moments." I laid down and instantly felt light headed. There was a very light hum, a rumble, and then nothing.

Chapter 6

 I woke up wondering where I was. Bright light shone through the windows. I scanned the control panel. Five and a half billion years ago! Are we lost?

 "Are you all right?"

 "Coming around." My head slowly emerged from a fog. "I appreciate the size and upgrades of this ship, but I'm five and a half billion years away. Did you overshoot?"

 "Never … you're right where you should be. Take a look, get oriented, and we'll watch Earth grow."

 I expected to see the formation of a budding ball, but as I walked around the ports, the view was more of an outer space wonder. Below, and to my right, a large cloud of material stretched as far as I could see and curved to the left in front of me. I looked behind and it was curving to the right. It was a tremendous ring encircling what I assumed was the sun directly to my left and below me. I took a few seconds to get used to having a sun below me. Its brightness allowed me to see only a fraction of the diffuse ring. With a simple entry on his input pad, Ralph engaged what appeared to be filters in the viewing ports to allow us to better distinguish objects in the bright backlighting. As I acclimated to the view, I could see that the ring was poorly defined. I was looking at a bulbous cloud nearby, but the cloud material thinned out towards the sun and then two ridges of increasing concentration appeared. It was more like a large plate with a rings of food spread upon it in concentric circles. I focused to the far right and could see hints of larger rings, further out from the sun.

 I looked to the left. "Venus and Mercury?" I asked myself, but Ralph answered.

 "As you call them."

As I studied this generalized ring pattern, objects streaking by from every direction and on paths of their own, constantly disrupted my focus. Some were dust-like and others many times the size of our ship. Some had comet-like tails and others were like giant rocks. There were diffuse belts of objects in orbits other than the plane of the major orbits.

"Oriented?" Ralph asked.

"Getting there. Am I looking at the sun and major belts of materials that will eventually become its planets?"

"And where are you?"

"Just above the plane of the planets and just inside Earth's orbit?"

"You're oriented; now let's take a closer look." We approached the third ring from the sun.

As we closed in, I focused on individual objects. They were of different sizes, shapes, and hues. Most of the pieces were moving in the same general direction, but their speeds differed and errant objects frequently streaked right through the parade of individual particles and bundles of agglomerates.

We were moving with the pack. I wondered how fast we were going and found the speed indicator fluttering around zero. Was speed relative to Earth?

"You're grasping it. Speed is relative to Earth and there isn't much Earth yet. Do you see it?"

I searched the field of dust and boulders, finally spotting a large object with two slightly smaller objects approaching it from behind. All three pieces were rough shaped and appeared to be a conglomeration of the space junk floating around. The smaller objects docked into the larger, rotated, and settled into comfortable positions.

"I'm watching the agglomeration of larger particles. Is that zygote looking bundle the budding Earth?"

"That's it. And this space junk floating by and the objects streaking through it contain all the minerals that will eventually make up Earth as you know it."

"It sure doesn't look like the boiling metal hell I envisioned. This looks like a very gentle process."

"So far, but hang on for a billion years."

Chapter 7

We went to an altitude of twenty thousand miles and sped through hundreds of million years of Earth's 'accretion' ... joining together ... of space particles. As time went on, other planetoids formed in Earth's neighborhood. It was like a contest of who could grow faster. With a head start, Earth's gravity was superior and it grew faster and faster as it took in competing planetoids. Incoming particles stopped cozying up to the growing Earth and began to whack into it soundly, penetrating the surface.

As time passed, the collisions got more violent and Earth's increasing gravitational force captured escaping volatile chemicals as a growing atmosphere. Approaching incoming objects then began to heat up in a display of fireworks as they penetrated Earth's gaseous blanket.

The atmosphere became murky with particles thrown up from the bombardments. As the atmosphere grew, we had to use more window filters to see through the haze, then different filters as wispy clouds developed in its upper reaches. Further growth resulted in an ominous and very dark cloud cover.

From our vantage point, it appeared that anything smaller than a house was vaporizing in the thick atmosphere surrounding the Earth. Larger particles bashed into the surface and caused the uprising of gasses and a flare of illuminated particles. I watched in awe. I wondered what was happening in the interior.

I should have remembered not to think. "What do you think is going on in Earth's interior?" Ralph asked.

Years ago Earth accreted in a rather gentle manner. Pieces of chemically different materials came together without the violence seen now. A piece of accreted material with its composition of elements could snuggle up next to a lump of material of a different history and composition in complete compatibility. I had seen lists of compositions of meteorites, comets, and planetoids. They were basically a variety of chemical compounds that would form into different compounds when they were heated together to their melting points, probably with the evolution of a lot of heat. We watched the Earth's surface bombarded with heat producing missiles, but what was happening down below, thousand of miles from sight?

"It has to be hot. Obviously molten or close to it."

"What are the sources of heat?"

"The kinetic energy of incoming mass is a major source, but chemical energy has to be a factor as a variety of incoming minerals melt to form free metals and other minerals."

"Don't forget radioactive decay," he added.

I had overlooked the fact that many of these young elements were highly radioactive and their decay process would create an enormous source of heat. "As the Earth grows, its gravitational force increases and its early softly packed structure compresses," I ventured. "This generates heat, and heat from the surface is transferred to the core as liquid material is pulled down by ever increasing gravity. This heat, and the heat from radioactive decay causes melting throughout Earth's deep core. The result is pure metal for Earth's core, a semi-solid magma, and lava slags that will eventually solidify to create Earth's rocks, and salts for its future oceans."

"Crude, but not bad," Ralph said. "Geologists may not appreciate your calling their future rocks slags, but your metal refinery analogy is quite appropriate. What do we have right now?"

I looked at the large ball below. In normal view, it was a dark muddy appearing spherical mass with orange blotches. "Our current Earth, a hot ball with a dense atmosphere."

"See anything different in the surroundings?"

I looked all around and far out into space. "It is much clearer, the agglomeration process has cleaned up the area.

"Look closer to home."

I looked at the neighboring materials. Everything appeared to be agglomerated from the original space dust and ranged from marble sized to 1% of Earth's diameter. In the distance, the largest of the particles I could see was bearing down upon Earth at an amazing speed. "Watch out," I shouted.

Ralph laughed. "We'll survive this."

The ominous object was on its own orbit. It streaked in and crashed into the forming earth at just above the mid-level of its northern hemisphere and well beyond the longitude of a direct hit. A little higher and it would have blown through the atmosphere, missing the planet.

Lack of a direct hit did not diminish the energy of the crash. This missile was a budding planet in its own right with its own developing atmosphere. It penetrated Earth's hot plastic surface and exploded. A small portion of mass, probably a mixture of Earth and the incoming object, continued to travel and started to curve ... captured in orbit by Earth's gravity.

The surge of venting gasses from the collision, and the capture of most of the smaller planet's atmosphere left Earth's expanded atmosphere in violent undulation. A trail of fiery and popping pieces followed the separated moon, cooled, and fell back through Earth's smoky haze.

"That was impressive." As we traveled through time I looked closely at the rotating Earth wobbling on its axis and then settling down, tilted from its perpendicular orientation to the sun.

"That is the formation of the moon, the adjustment of Earth's rotation, the tilt of its axis, and the setting of Earth's orbit around the sun." Ralph sounded proud. "But as you see every night you look at the moon's pockmarked surface, there's still a lot more action ahead."

"We've got the Earth day set with its spin, the year with its orbit, and the seasons with the tilt. What more do we need?"

"There are a lot of useful minerals out there and the planet has barely warmed up." Ralph sped up the time. At our rate of time travel, Earth and the moon were bombarded at a rapid rate, with the Earth getting the majority of hits and the largest particles. Energy flowed into Earth and it began to glow a bright red underneath its dense canopy of dirt filled atmosphere.

"Any hope for volatile materials staying in the interior of earth?"

Where was this guy leading me? Was he leading me? "Like what?"

"Like water, carbon dioxide, or nitrogen."

Not a prayer, I thought. No matter the chemical state of the incoming materials, the molten interior planet had been bombarded, mixed, stirred and purged of volatiles. Gasses are not compatible with molten metals and minerals. The refining fire of the planet was separating a mixture of space junk into liquid metal, slag, salts, and gasses. "Significant quantities of volatiles would not reside in the current Earth's interior," I reported with confidence. "This process is a very efficient separation of molten solids and gasses. It's just like our Earthly metal refining processes."

"Look closely at the equatorial surface."

I watched for a few seconds and suddenly an immense flare erupted out of nowhere. It looked like the reaction from the hit of a large deeply penetrating asteroid, but there was no preceding collision. "What was that?"

"You tell me," Ralph replied.

"Was that the release of gases entrapped in Earth's core? A belch like that would do a lot of mixing of the core and assure the removal of volatiles."

"Good answer, but you need to expand your thinking here. We are dealing with a much different scale and much different conditions than you had in your little chemical reactors."

A chilly silence filled the room. My answer was not good. This was an order to think if I had ever heard one. I must be missing something basic. I had long scoffed the earth scientists for their construct of an early cold and crusted Earth that was devoid of water and atmosphere . . . then released oceans and air. I shut the door on this idea. My Earth would not have gasses spewing up from its core for eons. Was I wrong? This was a different world. Could I find proof?

I felt the chill come out of the room as I began some open-minded thinking. There was a lot of Earth, and only a little skin of surface with a wisp of an atmosphere. There wouldn't have to be a lot of volatiles stuck in the core to make a real difference to the minor amount of water and atmosphere that is so important to us. The tremendous pressures at the core could cause a solubility of volatiles that would be unthinkable at surface conditions. That big belch we saw sure wasn't the last of it. Heating and the formation of gaseous byproducts from radioactive decay would create convection in the liquid core and move material around like gas in a bear's belly full of green apples.

"That's a great analogy." I heard Ralph laugh and slap his thigh. "Now, I have almost delivered you to your wish. We are nearing the time for Earth to cool, a crust to form and waters to cover it. When do you think that will happen?"

I looked at the time indicator.

"Not what time." Ralph sounded like an impatient instructor. "What conditions must exist before Earth will allow water to condense?"

I looked out onto the hot ball of Earth with its nearby moon circling it every few days. I was privileged to see this beautiful birth process unfold and overwhelmed with the rugged beauty. I knew that the end product was even more beautiful and wanted to get on with the show. Why doesn't he just tell me, I thought.

"I cannot make you smart, I'm limited to bringing you here and giving you the opportunity to observe and think."

I should have known, I stared at the impossibly hot Earth and thought out loud. "The surface is currently hotter than the critical point of water, so there will be no condensation for a while. What is water's critical point?" I mused.

"374.15 degrees Celsius." Ralph could not make me smart, but he could supply data like an onboard Wikipedia.

Ralph laughed at my thought. "I can supply you with certifiably correct and unbiased answers for anything you have known at one time in your life."

I thought more about the condensation process and at what point water would condense out of a mixture of gasses, but it all was irrelevant if this ball didn't start cooling. I wondered what Earth's surface looked like and reached out to toggle in a view-clearing filter.

"You're ready to take a closer look, but let's do it in normal view." I took my hand from the toggle. We weren't actually going to travel to Earth's molten surface. No way could we survive there, not even in this ship.

Chapter 8

"**WE'RE GOING DOWN THERE?**" I began to sweat at the thought. Nothing could survive at the current temperature of Earth's surface.

"We're safe," Ralph said without concern and completely ignoring my lack of faith and discomfort. "Give the atmosphere a little thought before you go into it. What do you expect to find?"

I took a deep breath. It certainly was the opportunity of a lifetime to learn about Earth's early atmosphere and we should go through this slowly. In my lifetime, I had heard many speculations on Earth's early atmosphere. They ranged from pure methane, to pure ammonia, and a wide variety of gaseous mixtures. Now I was there with an expert guide and a full range of analytical devices. What an opportunity. I gave Ralph my first impression.

"I certainly expect to find water vapor, carbon dioxide, methane, nitrogen, and the noble gasses ... all the components of the atmosphere I left in Cancun ... but what else?" I looked at the inferno below and recalled reading that the early Earth's atmosphere was void of oxygen. It certainly should be. There is no way oxygen could survive in this inferno. There was a lot of oxygen tied up in silicon and aluminum oxides in Earth's mantle and crust, but much of the world's remaining supply of oxygen had to be in the atmosphere tied up as water, carbon dioxide or carbon monoxide. "What would this mixture of gasses be?" My head was in overload as I attempted to recall all the chemistry of gases that I'd ever known or been exposed to. "The pressure down there! It is like nothing on Earth." I realized how stupid that sounded as soon as I mumbled it. I heard Ralph chuckle.

What happens at high temperature and pressure with a mix of gasses? I felt my head to see if it was overheated by the high speed combing of my memory bank. There was something simple back there ... something old and basic. A piece of my history hit me like a lightning bolt . . . the "Water

Gas" reaction. In the days of my youth, every town had a coal gasification plant that made and stored gas for home cooking. They were abandoned when natural gas pipelines came to town. My foggy memory was recalling a different version from Great Britain. What was that chemistry?

"From your country, coke was reacted with water to produce carbon monoxide and hydrogen. In Great Britain, the reaction of carbon monoxide and water to produce carbon dioxide and hydrogen had the same name," my on board Wikipedia supplied.

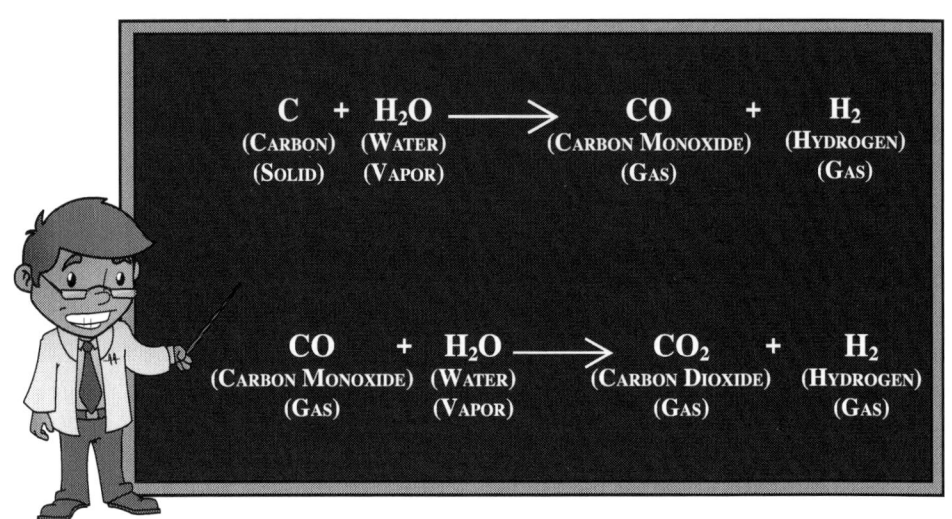

$$C + H_2O \longrightarrow CO + H_2$$
(Carbon) (Water) (Carbon Monoxide) (Hydrogen)
(Solid) (Vapor) (Gas) (Gas)

$$CO + H_2O \longrightarrow CO_2 + H_2$$
(Carbon Monoxide) (Water) (Carbon Dioxide) (Hydrogen)
(Gas) (Vapor) (Gas) (Gas)

Either way, any carbon in this atmosphere would be oxidized by water to its mono or dioxide and some water would be reduced to hydrogen. Some of the hydrogen would diffuse into outer space. With all the hydrogen and carbon monoxide in the atmosphere, it would take a long time and several miracles before we saw any free oxygen or ozone. This was not a friendly looking planet. Are we lost?

"You are right where you asked to be. You are looking at Earth just before liquid started condensing on its surface. Is it all right that I took you back a little further than you asked?"

I looked at my starting point. If I had been dropped here without seeing it form, I would have had millions of questions about its origin. "Thanks, without being here I would have been overwhelmed by the next step into the future. Now I'm just plain scared."

"You have nothing to fear, you are ready. Take the controls and have a close look."

My fears of Earth's hellish conditions subsided and were replaced with a concern for sufficient time to see what I needed to see. I felt like I was entering a major gallery or museum and knew I would not see everything unless I hurried and if I hurried, I'd appreciate nothing I saw. I wondered how much time remained on the clock still ticking away on my sleeping wife in Cancun. "I know I don't have forever on this trip." Have we used up much of the Cancun night yet?"

"The Cancun night is barely started. If you were worried about taking time for a look around, no te preocupes, if you stick to your agenda you have plenty of time."

"I assume the same rules that guided my test drive of this morning apply ... I do what I want to do and you protect me. Right?" I struggled to sound confident. I was sweating and having to blink my eyes to clear the light-headedness. No way could we survive this hothouse.

"Trust me amigo." His humor supplied sufficient confidence to grab the controls and head toward Earth's molten surface.

I scanned the control panel and thought about plugging in some constraints. I wanted a close look, but not a static picture. I responded to the "Real Time" option under the time menu and pushed the joystick down. As I approached the ominous black ball tinted with reddish blotches of hot spots and speckled with wisps of white puffing above the surface, the phrase of a religious creed "descended into hell" crossed my mind. What was I getting into? I hesitated.

I had grown up with the comfort that a loving god had created the world in six days. As the fascinating discoveries of Earth's cosmic history unfolded during my lifetime, I joined with those who celebrated the continuing presence of a larger and more loving god rather than fall into the trap of choosing between an ancient concept of god or no god. This was not the first time in history that Judeo-Christian beliefs were challenged by science, there were brutal arguments when the world became round and even more when it was no longer the center of the universe. Now I was witnessing early Earth formation. Was this possible? Why did I deserve this opportunity? Or, had I earned a punishment?

"Go ahead," Ralph urged. "You've always been in god's hands."

"Thanks for the reminder." I pushed the stick down.

Going through the clouds was similar to descending through thin, icy, high altitude clouds on any plane flight. Looking back through them at the sun was pretty much like home.

The atmospheric composition readouts were from another world. The dominant gasses were water, carbon dioxide, methane, carbon monoxide, nitrogen, and hydrogen. The rare atmosphere I entered sat on top of a thick layer of gasses that contained most of Earth's carbon, oxygen and hydrogen plus the water to fill the oceans. I pushed the joystick down to descend into the deep dark layer below.

The spacecraft was supposed to be immune from shaking, so I guess that it must have been me that shook when I broke through the surface of what proved to be the tops of angry clouds … very angry clouds. We entered the cathedral of a ferocious lightning storm, magnitudes worse than anything I had ever seen. I dove below this band of clouds, broke free of the constant jolts of electricity, and looked up at a dancing pattern of cloud-to-cloud lightening. I was in a downpour of epic proportions. I cruised at a constant altitude and in a straight line, thinking that all this flow could not be in one direction. Sure enough, I hit a rising thermal. The rocketing column of gas carried golf ball sized "dust" on a vertical path. The mottled red and black appearance of Earth from space now made sense. The rising thermals were warm and filled with hot debris while the thick cloud tops hiding the lightning were black with precipitation.

I didn't care to search through hundreds of miles of rain, so I plugged 500 feet into the altimeter and punched the "GO TO" button.

"Bored?" Ralph asked.

"No, would I have missed anything in a hundred miles of rain?"

"Not really, but I'll bring you down slowly. Take a look at what happens to a rain drop on the way down."

I had not given that a lot of thought. In my Cancun world, raindrops agglomerated at high altitudes and fell. If they went too fast, they probably broke up. What would happen in an environment where rain fell through miles of ever increasing density? And what was this rain? It certainly was more than water. With this atmosphere, it had to be very acidic.

Acidic, I really hadn't thought of where the chlorine, bromine, and fluorine would be. In this reducing atmosphere, would the halides be present as hydrochloric, hydrobromic, and hydrofluoric acids, all very strong acids?

"I can answer that."

Would I get a direct answer and not be forced through an arduous thought process? This was a pleasant shock.

"I can provide the miscellaneous background information, but you'll have to work to get carbon information. Yes, halogens would be present here as their strong acids, but they had better work to do and have been consumed by the reactive metals."

I had not considered the reactive metals. Calcium, magnesium, sodium and potassium would react very strongly with the halides to produce ionic salts that would soon fill the oceans. Where was the sulfur? It probably was also present in its most reduced form as hydrogen sulfide.

"Much of Earth's sulfur is out there as hydrogen sulfide." Ralph nodded toward the window.

I looked outside and shuddered, dropping back in my seat.

"Scared?"

"Not really," I lied, "but what a poisonous mess that is outside. I wouldn't last a minute there."

"We've got more to learn. Let's go down."

I had more of a feeling of going under water than descending through air. My heart was in my throat. I took several deep breaths. I was sweating. The "rain" drops were growing to football size, but their rate of descent seemed to be slowing. The density of the gas we were descending through was approaching that of the liquid falling. The giant liquid drops streaked out and it became difficult to tell gas from liquid. Earth's surface came into view as the spacecraft eased into the requested 500-foot altitude.

I was right in the region of vaporization of the falling liquid. Earth was well above the critical temperature of water, the temperature above which water will not condense at any pressure. The liquid streaks were being evaporated by heat radiating from Earth's molten surface. Because there is little heat of vaporization or change in density near the critical point, the transformation of liquid globs to vapor was a rapid process. The liquid droplets vaporized and their entrained particulates splattered into the molten yellow lava surface. The larger ones stayed dark for a few seconds before being melted from below.

I wondered how it would be best to view the development of the oceans. Should I stay here and call for time to move on, or would the view from above be better?

"You'd better stay here and think for a moment," Ralph said. "Where will the halide salts be now?"

Chapter 9

I ROAMED BETWEEN THE HELL of Earth's molten surface and the waters above that were held at bay until more cooling occurred. I wondered how long this craft could remain viable in this intense heat? How long before we were cooked alive? My head was too overloaded with fear to address Ralph's last challenge. Where were the halide salts?

Halides salts, such as calcium chloride and sodium chloride that would soon be dissolved in the oceans in tremendous quantities, would be extremely stable and must have formed in Earth's early chemistry. They should be present on Earth now. Would they be mixed in the magma of lava? Probably not; they are physically and chemically much different from lava. My brain was overloaded and I started to worry about being cooked alive.

Sweating increased and I was beginning to feel weak and light-headed. I'd experienced these feelings before, usually induced by too long in a hot tub, dehydration, or low blood sugar. I would go into a rapid but futile heartbeat that pumped little blood. Only once did I require emergency room intervention, and after that a little bit of Toporol has kept me in control. Had I gone out of control? I looked out the window and had difficulty focusing. I felt for my pulse at my throat and could feel nothing but cold sweat.

"Get to the couch!" Ralph was out of his chair and helping me out of mine.

He sat me on the couch, gave me a glass of water, and then laid me on my back. He put his left hand on my throat. His eyes developed a mysterious glow and he tapped my rib cage above my heart with his right hand. I felt immediate relief.

"You'll be fine." He brought me back to the controls.

I felt much better, far better than any time on the trip. Who was this Earth Keeper? What was his role? I'd better just get back to work and be grateful, but at some time I hoped to find out more about him.

I slowly cruised under the impending rain, feeling like I was on a hot skillet just under dancing water drops. It seemed impossible that any ship could function under these conditions, but my faith was restored. I moved horizontally with the flow of liquid above me and looked closely at the lava below. Lava melts at around 500 degrees Celsius, well above the critical point of water, so there was going to have to be a good crust on the Earth before any water condensed. Current temperature was over 500 degrees and normal Earth lava would be molten, but by looking carefully, I could see islands of solids.

Were they salts islands? Salt is ionic and chemically much different from molten rock … they should not be expected to dissolve in each other. Did halide salts have a higher melting point than lava?

"Melting points range from 600 to 800 degrees. Boiling points from 1300 to 1600 degrees. They are as stable as rocks."

This was amazing. In the millions of years that Earth was a hot ball with no liquid water, stable halide salts formed and grew into floating islands on Earth's surface.

I scanned the horizon in all directions, looking closely at the surface. Islands of solids in the oceans of lava were becoming apparent. In my thinking, as Earth cooled, halide salts would solidify first and float on top of the silicon and aluminum oxide rich molten lava. I assumed they would be lighter and float, but I'd better check.

"What's the density of lava and solid sodium chloride?" I queried my onboard Oracle.

"Two-point-six and two-point-one-seven respectively," came the immediate reply.

"Then salt will float on slag. Are those islands of halide salts floating on the lava?"

"You're learning. What's your current picture of Earth?"

"I have an Earth that is the product of a metal refinery. Some heavy metals were refined to the core. Slags that will become rocks now have halides that will become salts of the ocean floating on top of them." I took Ralph's lack of reply as confirmation.

I turned left and traveled perpendicular to the streaking liquid above. Ralph told me how the actions of the moon tide and the weather systems moved isles of sodium chloride, the highest melting salt, around and caused them to agglomerate and grow. They were like icebergs with most of their mass below the surface. As sodium chloride became depleted and the temperature cooled, other halide salt islands formed. Most would eventually be dissolved in the ocean's waters when it became cool enough for liquid water to exist.

What a lesson! I had dozens of questions, but realized I was getting out of my field. It was difficult to relate to Earth covered with a liquid other than water and anything other than ice naturally floating on that liquid. I had seen monstrous piling of wind driven thin sheets of ice along the shores of Lake Superior and the stacking of pressure ridges of thick Arctic ice, but nothing like these salt island creations. Our lakes and oceans were subjected to annual seasons … while this process of stacking up solids continued for millennia.

Thinking of the Great Lakes reminded me of recent comments from a friend who bemoaned the fact that twenty years ago he could drive far out onto Lake Erie to fish for perch. Now, only a rim of ice encircled the lake in the middle of winter. Living in the "lake effect snow" shadow of Lake Michigan certainly allowed us to feel the effects of warming. Years ago, the snows diminished when the lake froze over. Now we have lake effect snow all winter long.

"Feeling better?" Ralph asked.

I was, and realized I was so comfortable that my mind had peacefully wandered. Back to work.

The overhead streaking curved to the right. We were moving toward the center of a cyclonic system of unearthly proportions. The high winds and dense atmosphere pushed large slabs of salt into each other and stacked them up into windrows. The uplifting cyclone forces and surging moon tide created a chain of islands arcing out towards the horizon. The lee sides of the larger islands, protected from bombardment by dirt particles, were covered with spectacular crystalline salt formations. Globs of lava the size of basketballs were ripped from the turbulent surface and sent spiraling skyward.

"Satisfied with the pre-water Earth?"

I had millions of questions. I would have loved to know where the gold, silver, platinum and other valuable minerals were and what their fate would be, but that really wasn't any of my business. "Immensely, thank you," I replied and went straight up the rising flow of the cyclonic eye.

"I'm glad you're satisfied with the geology, now let's think about the chemistry."

"Good enough for me …" Crash! A bolt of horizontal lightening flew by and left a streak as far as I could see. "I wonder what chemistry is going on in that?" I thought out loud.

Chapter 10

   Ralph pointed toward the crackling lightening bolt. "What kind of chemistry do you think is going on there?"

   "I know of experiments with synthetic atmospheres and electrical discharges producing all manner of chemicals in small quantities, even some complex amino acids. Many people think that life emanated from an early lightening storm."

   "Stick with gross chemistry," was his rather testy response. "You will learn more if you stay with what you know."

   What would be more interesting or enlightening than learning about the full spectrum of chemicals that could be formed in this soup of a high-pressure reactor with lightening as a catalyst? I guess that Ralph didn't want me wandering off into the origin of life and there must be something big right under my nose. What could it be? ZAP ... another monster bolt of lightening streaked by.

   "Wow!" I shouted. "That thing would fix enough nitrogen to fertilize Earth for a month."

   "Really?" His tone was assuring. I knew I had to go further.

   "On my Earth, lightning fixes tens-of-millions of tons-per-year of nitrogen into ammonia. Would it do that in this atmosphere? It certainly is loaded with hydrogen and nitrogen, the building blocks for ammonia."

   "If it did form ammonia, what would be the ammonia's fate in this soup?"

   From his tone, I felt that he was pushing me to think back ... way back. I remembered a college chemistry study on the manufacture of urea, the commercial fertilizer. It was formed from the reaction between ammonia and carbon dioxide under high pressure. Was that the condition I was observing? Would ammonia eventually be scavenged out of the atmosphere as urea and end up in the developing ocean?

"The developing ocean is going to have a lot of nutrients." I hoped I was right and waited for Ralph's assurance.

"Ironic isn't it?"

"How so?" I asked.

"In 1828 the laboratory manufacture of urea was celebrated as the first "organic" synthesis, the first manufacture of a compound previously made only by living organisms. Now we are looking at megatons of it being made above the budding Earth well before the presence of any living organisms."

I was flabbergasted at the prospect of the Earth covered with a high nutrient ocean and eager to watch its development.

"Let's not worry about nutrients for a while," Ralph said. "The ocean is also going to load up on minerals. Let's let it rain for a few thousand years, before we go down for another look."

Chapter 11

WHILE WAITING FOR THE OCEANS to fill, I thought it might be a good time to find out more about Ralph. He seemed to be opening up. Maybe I could probe. Why not ask a direct question in a calm manner like I'd ask someone; 'What time is lunch?' I decided to go for it. He wasn't going to throw me off the ship.

"Ralph, just what's in the job description of an Earth Keeper?" I asked without even looking in his direction.

"The main goal is to end up with a planet having a sustainable non-warring population of intelligent life," he answered as matter of factly as I asked.

Wow! He has that job? And this ship? Who is he working for?

"I'm a servant of the mother of all universes." He answered my thoughts before I could ask.

What's Ralph's motivation, I wondered. Does he want to help us? Is his success rewarded? Failure punished?

"It's a competitive job. If I do well on Earth, I'll get a bigger challenge for the next cycle."

"What's a cycle?" I blurted out without any pre thought.

"Your science now understands that Earth started out as a dense mass and has been expanding since a "Big Bang." When it stops expanding, a half a cycle is completed."

"So our universe will return to a dense point and you will get another assignment? After the next big bang and expansion of matter into energy and then back into human friendly planetary matter?"

"Or in a larger or smaller universe, depending on how well I do this time."

"Other universes? Where? Was this your first assignment?" I was blown away with his answers and talking faster than I was thinking.

"This is my first time in this universe. I was promoted to it from past successes. As to where the universes are, some of them overlap with your universe . . . and others are separate from it."

I had a hundred more questions, but several thousand years of rain had fallen and it was time to catch up. "Back to work," Ralph said with no particular inflection and we returned to the pursuit of Earth's early development. "This is an interesting period. It is still above 250 degrees Celsius (482 degrees Fahrenheit.) Only about thirty percent of the atmospheric water has condensed."

I looked over areas of surprisingly rugged mountains created by buckling and stacking of Earth's thin crust. Violent tides washed over all but the most elevated of land areas. We circled at astronaut altitude. The high ground, buckled up from below, was leaking out molten lava and venting gasses.

We went up in altitude and increased the speed through time to 5000-years-per-minute and faster. At these speeds, Earth was actively building itself as it grew oceans and land in a kaleidoscope of patterns. Proto-continents rose, fell, and skittered about Earth's surface. Relentless rain scoured the land surfaces, eroding them and dissolving the exposed salts. I could have watched this for hours.

The rains continued and the oceans grew. Landmasses developed higher and oceans deeper. With time, urea, ammonia, and other chemicals formed complexes with and dissolved copper, nickel, iron and probably dozens of other metals to give the "water" an eerie blue green color.

"You're getting it." Ralph sounded fatherly, proud of me. I sat back in amazement. What a mental overload. It was unbelievable; I was looking at the beginning of oceans on Earth's hot surface. I pinched myself.

"Don't hurt yourself," Ralph said. "Let's get moving. There isn't going to be a lot of excitement in the next few thousands of years of rain."

Chapter 12

    RALPH TOOK CONTROL AND WE flew through time, watching the composition of the atmosphere and its appearance change from a hundred miles above its outer limits. We stayed at a constant altitude and with time whistling by we could see the ball of atmosphere shrink as water vapor and carbon dioxide transferred to the cooling ocean. The amount of debris sent up by bombardment from asteroids and volcanic emissions slowed and the constant rain scrubbed particulates from the atmosphere. Unrelenting lightening filled the growing oceans with nutrients.

    Under modern conditions, with the presence of life, a little more cooling would turn this ocean into a cesspool of slimy living things. But now there was no possibility of life in these harsh conditions. Most of the strong acids must have been scrubbed from the atmosphere, but what was the current atmospheric composition? What was the temperature?

    The "Analytics" display lit up. The atmospheric pressure was down to sixty percent of the starting pressure. There was an abundance of carbon monoxide, methane and hydrogen, but no oxygen. The ocean temperature was still 150 degrees Celsius (302 degrees Fahrenheit.) The hydrogen sulfide content of the atmosphere was still high.

    "No place for life down there," I muttered.

    "Not life as you know it, but let's look around." He showed me living fronds waving in the serenity of a calm bay and assured me that there was much more life in the deeper waters. "The life forms your scientists are discovering near oceanic thermal vents are descendents of this current outgrowth," he assured me.

    Ralph sped up the clock and we cruised over the ocean at about a 50-mile altitude. We stayed out of the rainiest areas and maintained a general overall view. At the speed we were going through time, isolated parts of the ocean turned through a kaleidoscope of colors; green, orange, red, blue, black and yellow.

"That is spectacular," I said, "But what is it?"

"We are cool enough to take the hydrogen sulfide out of the air …"

"And react it with heavy metals to produce those spectacular mineral deposits," I added.

"Right. Now what's remaining to take care of to get to life as you know it?"

"That's easy, get rid of this atmospheric carbon and get the temperature down."

"And how are we going to do that?"

"Bury it as coal and oil … dead plants and dinosaurs … just like the plot goes."

"Do you see a better choice, a very large biological reactor?"

"The nitrogen filled ocean!" I jumped out of my chair, thinking of another critical element. "All the world's supply of phosphorous must be in there too. If you supply that with the right life, it will go absolutely wild."

"Wilder than you can imagine." Ralph laughed and we started rushing through time.

Chapter 13

WE WENT ON FAST-FORWARD TO watch the oceans grow and cool. The atmospheric pressure was down to a third of what we had started out with. A lot of water and hydrogen sulfide had been wrung out, but it was still heavy in methane, carbon dioxide, carbon monoxide, nitrogen and hydrogen as the temperature went through 100 degrees Celsius (212 degrees Fahrenheit) and continued to fall. I watched the ocean for blooms of life and was hoping to see some green stuff that would take carbon out of the atmosphere and pump oxygen into it.

"Don't worry about us not cooling down," Ralph said. "Remember, the sun is not as strong as it will be in the twenty first century."

I'd completely forgotten that the sun, and all stars in the sun's current cycle of life, got brighter with age. We were still in the warming part of the sun's cycle. Maybe this was a good chance to check out a concern with the modern sun. By the best satellite measurements taken over recent decades, Earth was receiving a variable amount of energy from the sun. It was currently at a low point and seeming to stay there. Would it? Were modern levels of greenhouse gasses affecting temperature as much as this natural variation? I wished I could see the graphic I had in my folder in the condo.

Without asking, my very graphic appeared on the display screen. The modern sun had an eleven-year cycle and its energy variation was regular, but in looking carefully at the numbers, energy output varied only a tenth of a percent or so. How did this compare with the predicted effects of increased carbon dioxide levels? Would this variation continue into the future? Was this modern variation going on all through the sun's history?

"This variation has gone on a long time, and will continue," Ralph said, "It is only a small fraction of the effect of the increased retained energy coming from the higher levels of carbon dioxide. It may have helped trigger an ice age in the past, but it will not turn around the human induced warming of the 21st century"

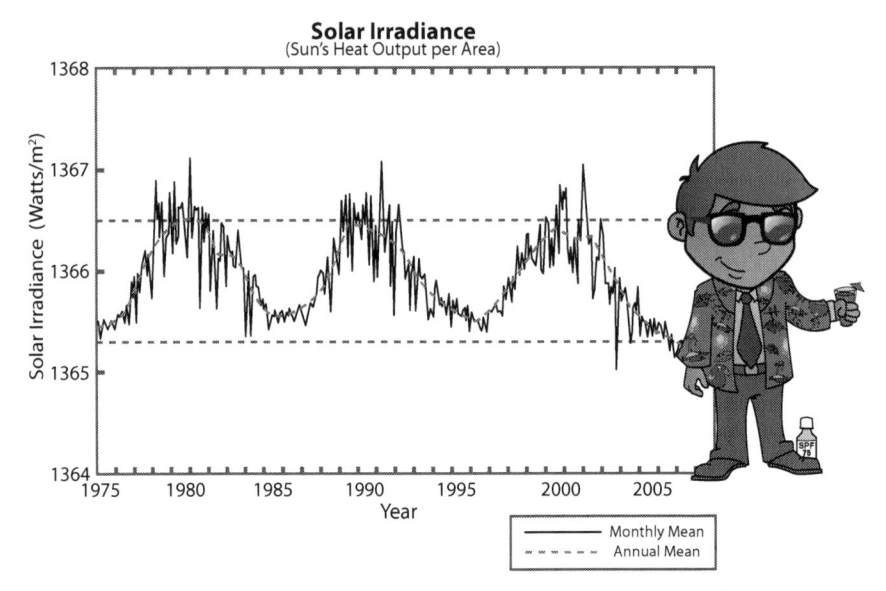

Solar Irradiance
(Sun's Heat Output per Area)

[Graph showing Solar Irradiance (Watts/m²) from 1975 to 2005, ranging from 1364 to 1368, with Monthly Mean and Annual Mean lines]

We flew through time while staying in the lee of a landmass, a place where phytoplankton or other green life could be seen in the calm waters. Finally, greenish clouds of filamentous material began showing up, replacing their predecessors who had gotten rid of the hydrogen sulfide at higher temperatures.

"That's amazing," I mused, "a new life form taking over."

"There's a new job to do," Ralph said. "This one is super-efficient at reproducing, growing, and producing oxygen.

They certainly were. We cruised through a few thousand years and the ocean turned green with life. Atmospheric oxygen appeared at very low levels, but did not increase while carbon dioxide levels plunged. The thin green pea soup of an ocean was rapidly consuming the carbon dioxide, but where was all the oxygen going? It wasn't coming into the atmosphere.

I saw spots of ocean that were brownish green, and then some that were reddish brown and got my answer. Dissolved iron was gobbling up all the oxygen the ocean was producing. Iron was dominant, but all the metal salts dissolved in the hot brew of acidic solvent that lashed the Earth under a reducing atmosphere were now exposed to oxygen for the first time. It took many years for the iron and other metallic oxides to form and deposit on the ocean floor. The ocean temperature dropped and it turned green with life as more forms of oxygen producers took advantage of the increasing sunlight, ideal nutrients, and friendly temperatures. It seemed as though it was cooling very fast. Were we in for a freeze?

"The rate of cooling may not be as bad as you think. We've been doing some wild time travel. Take a look at the graphics."

The display lit up and I asked for a temperature history. It was true; the rate of temperature reduction was slowing, almost stalling. I checked the graph of carbon dioxide concentration. As expected, it decreased slowly as the ocean life began, and then developed a rather steady rate of reduction until it too was slowing.

"Are we approaching a stable point?" I asked.

"No, just a slow period. We have to get rid of carbon dioxide and make a lot of oxygen yet. Have you looked at methane?"

Why? I didn't see how methane fit into this. Its graphic came up and I saw that the rate of methane decrese had slowed to nearly a stop. Why?

"Tough one, isn't it? I'd suggest you go down and check out the micro analytics."

I went to the surface of a bay, out of the wind, to measure surface gases. Methane was slightly high. The water was venting gas. Nothing more than I would have expected from the biodegradation of plant life in shallow waters, certainly not enough to cause the change I saw on the graphic. I was perplexed as I pondered the data.

"Try measuring over open water," Ralph offered.

Why, I thought. Methane hydrate is stable in deep water. I know a hint when I hear one, so I went out to sea and measured methane at the surface. The reading went off the scale I had set. My shocked reaction was rewarded by a chuckle. I was missing something and I was not going to beg for clues. I was embarrassed when I finally thought about the temperature. The ocean floor was well over 70 degrees Fahrenheit and methane hydrate would not form. Decaying vegetation would vent methane just like it does in shallow water. But why was there such a high rate of methane formation from the floor of the deep ocean? It was silent for a few seconds.

"I've got to help you here." Ralph sounded happy and positive, like he was thrilled to help. If he wanted to break his rules about not helping me with carbon information, I'd gladly take it.

"In looking up methane hydrate, did you notice that there were two ways that it is formed?"

I did seem to remember a thermal geological process and the biological process. I of course studied the biological process and ignored the geological process. "I believe there were bio and thermal routes."

"We are seeing thermogenesis now. During the past eons, vegetative matter has accumulated on the ocean floor and then been covered with a layer of volcanic ash, sediment, or whatever. Many layers have built up and the lower vegetative layers are now compressed and hot. These conditions form oil and methane that gravity brings toward the surface. If the methane does not hit a dense confining layer, it vents from the ocean floor. Now, the ocean's vegetative system is fighting to remove residual carbon from the atmosphere, as well as the recycled methane it had sequestered years ago. This will make for slow progress."

Progress was indeed slow. As we sped through time I noticed that there were fewer incoming and cooling continued. When the temperature dropped to below a hundred degrees Fahrenheit there was a sudden increase in the rate of carbon dioxide loss from the atmosphere.

"What is causing that?" I asked.

"Not this time," he said. "This one is for you to figure out."

I was looking at an area just inside the point of the bay where the current swept the water clear of biomass. Its sparkling clarity caught my attention. Ocean water had been cloudy with fine particles of calcium and other carbonates. Where had they gone and did that have anything to do with the sudden uptake of carbon dioxide?

It hit me like a ton of bricks. "We are at the temperature where dissolved calcium bicarbonate is stable. The powdery suspension is taking up carbon dioxide just like rainwater dissolving limestone."

"Only faster," Ralph assured me. "There will be a major rush of carbon from the atmosphere and into the ocean. The increase in acidity in colder water will kill off much of this biomass, and it will be replaced by strains that will function in new environment."

I appreciated the biological information. I would have had difficulty figuring out that sequence. The temperature seemed to be dropping uncontrollably. "It's going to get cold." I hurried down to the surface to check the temperature.

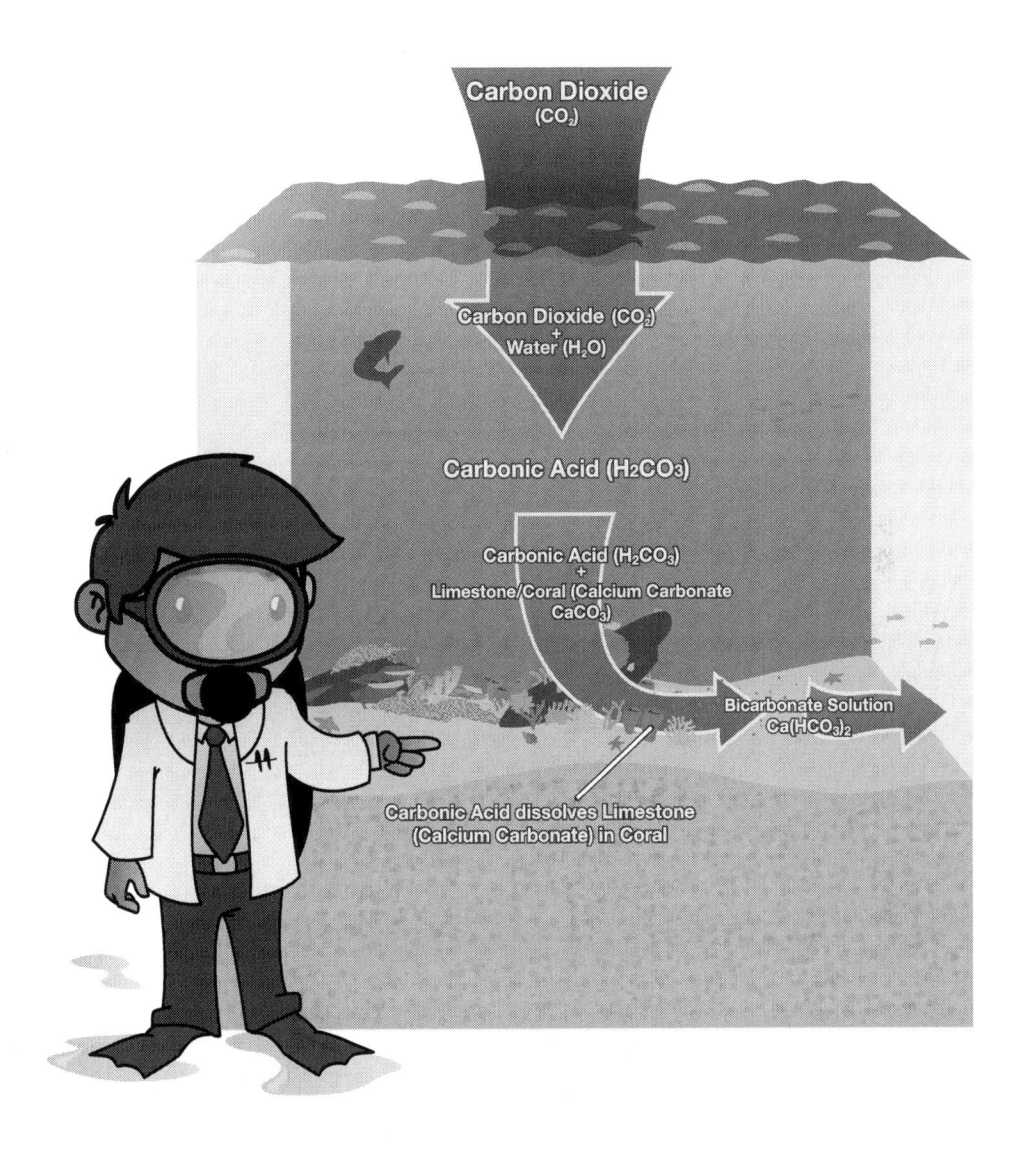

Carbon Dioxide (CO₂)

Carbon Dioxide (CO₂) + Water (H₂O)

Carbonic Acid (H₂CO₃)

Carbonic Acid (H₂CO₃) + Limestone/Coral (Calcium Carbonate CaCO₃)

Bicarbonate Solution Ca(HCO₃)₂

Carbonic Acid dissolves Limestone (Calcium Carbonate) in Coral

$$CO_2 + CaCO_3 + H_2O \longrightarrow Ca(HCO_3)_2$$

(Carbon Dioxide) (Calcium Carbonate) (Water) (Calcium Bicarbonate)
(Gas) (Solid) (Liquid) (Dissolved)

Chapter 14

THE SURFACE OF THE OCEAN was a mere thirty-seven degrees Celsius … human body temperature. "Life as I know it could survive here!" I shouted out that amazing fact.

"It has been surviving for quite some time," the calm and collected Ralph replied.

"But there is no oxygen," I protested.

"Take a look at the analytics history. Oxygen is trying to make its presence known. It just has too much work to do to establish a significant atmospheric concentration."

I called up the atmospheric history analytics. Water vapor pressure started plunging when the oceans began forming. Carbon dioxide levels fell since the beginning of biological life, but oxygen levels were at the baseline. Carbon monoxide was nearly depleted from the atmosphere and methane levels were falling slowly.

I magnified the scale for a closer look at low levels of oxygen. Sure enough, when the green oceanic mat started to build, oxygen levels crept above baseline. Carbon monoxide was readily oxidized to carbon dioxide in the lightening filled atmosphere. When carbon monoxide was depleted, the oxygen level rose a bit.

I felt as though someone was looking over my shoulder as I examined the analytics. "You're on track. Now what happens?"

"The ocean's plant life will continue to remove carbon dioxide," I ventured. "As more carbon dioxide is taken up by the carbonate powders, it will get cooler, then …Oh my God!"

"What?"

"When the deep ocean floors cool below seventy degrees Fahrenheit, the methane seeping up from lower layers will be trapped in the upper layers as methane hydrate."

"And?"

"The ocean will strip the atmosphere of carbon. With this weak sun, we are in for a freeze."

"I don't understand your fear," Ralph said. "You know that Earth will survive ... you are a product of its future."

"I know that it will survive, but if it turns into a bright white reflective ice cube, even the sun I left behind will not melt it. I don't know HOW it will turn around and warm up."

"And isn't that the reason you wanted to take this trip? To see how the Earth's atmosphere evolved and how it participated in recovery from ice ages."

"I guess so." I realized how silly I must have sounded. My fear of not understanding something was surfacing during a mission to gain just that understanding. I was suffering from information overload. Everything was happening so fast. Could I see an instant replay or something?

I heard Ralph chuckle and the display screen showed two simple diagrams of a cross section of the ocean bottom with layers of organic and inorganic deposits under the ocean floor.

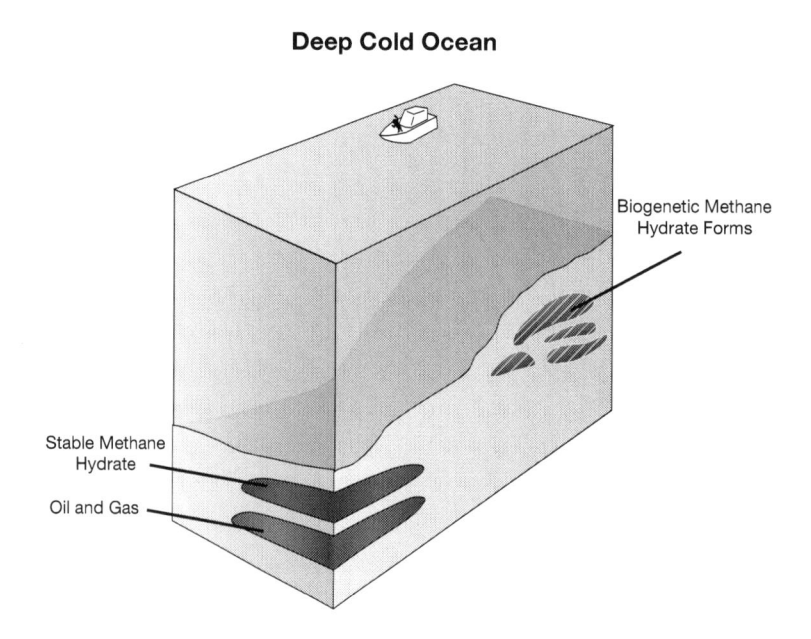

Deep Cold Ocean

Biogenetic Methane Hydrate Forms

Stable Methane Hydrate

Oil and Gas

Shallow Warm Ocean

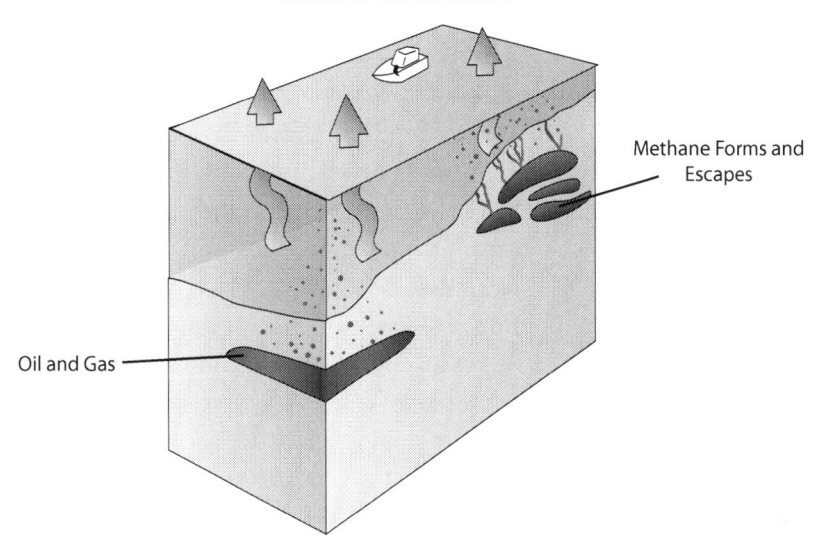

Methane Forms and Escapes

Oil and Gas

In the one titled 'warm ocean,' methane generated in the lower and warmer areas was bubbling up through the ocean and into the atmosphere. In the one titled 'cold ocean' there was no bubbling up and the methane was being trapped as methane hydrate in the upper layers of the ocean floor.

It was clear to me, and I did not wait for him to explain. "What a difference! As soon as the ocean gets cool enough to trap methane venting from deeply buried decaying vegetation, this up flow of methane is cut off. The ocean's ability to send oxygen into the atmosphere to oxidize methane will be greatly improved. Atmospheric methane will rapidly diminish.

"You got it."

"We are on a path to an ice cube? Will I be able to understand how Earth recovers from a major chill?"

"Hang on, it's about to get cold."

I remembered hearing about maverick scientists who postulated that the Earth went through a deep ice age, or ages, in its early history. They claimed to have evidence of glacial scouring on Australia when Australia was not far from its current latitude.

That reminded me. What were Earth's current continents like? Were plates moving? "Let's go up for an overview." I punched in three-thousand-miles altitude and started a pole-to-pole orbit.

More surprising to me than the impending cooling was the rate at which continents were forming, splitting and moving about the planet. At one time, there was one dominant continent with a spotting of shallow seas. The large mass split and other continents formed, grew, split, and crashed into each other to raise mountain ranges. Everything was propelled by volcanic activity at tectonic plate edges.

At most any time a violent volcanic reaction and its tailing plume of dirt and vapor was visible somewhere on Earth. At the speed we were traveling through time, asteroid collisions were still frequent and spectacular. The aftershock of large strikes at key points often altered the drift of continental plates. The continents were spread around various latitudes when we stared the pole-to-pole orbiting, but now they were lining up around the tropics. The globe looked like an ever-shifting puzzle that an artist from somewhere was trying to order to his liking. We continued a polar orbit through time and watched the continents ring Earth in a tropical band.

"This continental lineup is pretty stable," Ralph said. "It also reduces the ocean currents ability to transfer heat towards the poles."

We saw ice in the southern polar ocean on our next pass. This was a sure sign that the path to an icy Earth was straight ahead. Whitening Earth's surface would reflect the sun's heat and further tip the balance toward cold. We were heading to a condition that I could not imagine a way out of, Ice Cube Earth.

Early Freezing

Chapter 15

THE SOUTHERN POLAR ICE WAS lost during its first summer, but within a few centuries permanent polar ice caps reached the Arctic and Antarctic circles. This capability to make polar ice was similar to today's, but with the methane diminished, a weaker sun, and the carbon dioxide still falling, we were in for a real freeze. Gradually a stable platform of crushed and frozen ice covered Earth's polar caps to mid latitudes.

Earth's temperature plummeted as reflectance from the vast expanse of snow-white surface reduced the sun's net incoming radiation. The ocean ice cap raced toward the continents belting the tropics. Snow blanketed the high reaches of the jagged young outcroppings of folded crust and cold volcanic peaks. It was not long before the continental mountains were glaciated, but the shorelines remained ice-free.

I looked at the analytics. The atmosphere had stabilized and the sea level temperature at the continental edges was well above freezing. There was still a lot of open ocean supplying snowstorms of ferocious intensity. Would the ocean eventually freeze over? What could possibly turn this around?

"How will we ever get out of this?" I asked. Of course, I knew we would, but could not see the path.

"If it is any comfort, I can assure you that a large band of ocean will not freeze. We are in a stable thermal condition."

That was not much comfort, but it did cut down on a lot of recovery options. "So, will we just sit here and glaciate?"

"Until it gets warmer."

"And how will it get warmer?"

"What does it take to get warmer," he teased.

"The sun is getting warmer."

"We have to move faster than that."

"Then we are stuck with getting the sun to be more effective by decreasing the snow's reflectance or unleashing spectacular volcanic venting without spewing sun blocking particles into the stratosphere."

"Good thoughts," he said, "but there is an even more effective option."

So we had to get some global warming gas into the atmosphere. Carbon dioxide and methane were the obvious candidates. It was doubtful that we could get any carbon dioxide out of the cold and stable ocean. It must be methane.

"And?"

Yeah, and what? The oceans had been cool enough to capture seeping thermogenic methane as methane hydrate in strata below the ocean's floor, but how would we get it into play? There had to be an increase of temperature or a decrease of pressure to break the methane hydrate and allow methane to bubble to the surface. Temperature was not going to rise. A decrease in the ocean levels would decrease pressure, but the vast amount of sea ice did nothing to lower the level. We needed a lot of ice on land to lower the ocean, and it didn't look like this tropical band of continents was supplying it.

We continued to watch the glaciation thicken and calve icebergs into open oceans. Miles thick glaciers smothered vast areas, but ocean levels seemed almost static. Ocean levels dropped, but nothing like the lowering of the Gulf of Mexico that I had seen on the test trip. It didn't look like enough of a drop to release significant amounts of methane hydrate.

"If you want to decompose methane hydrate on the ocean floor by changing the pressure, and cannot lower the sea level enough, what is the other option?"

I felt like shouting "raising the floor dummy," how ridiculous could this guy be?

"Think." His voice was stern. I'd hit a nerve ... or made a lucky guess ... or made a fool of myself.

I thought, and the middle option felt most fitting. The ice caps from the last modern ice age had sunk northern land surfaces hundreds of feet and they slowly rebounded for millennia after the ice melted. Could this equatorial loading of continents with ice squeeze the Earth to raise its polar floors? Why not? No wonder the ocean level did not appear to be lowering. The continents, resting on a solid crust were carrying a heavy ice load and sinking into the magma below. If something sunk, something else had to rise.

"Good thinking. Let's go see." We headed to the South Pole and got there just as the ice cap was being shattered by a gaseous eruption from below. Fires spread across an area covering thousands of square miles as plate lifting and cracking, and the resulting volcanism, created a blowing and rolling mass of molten lava surrounded by an ever-expanding ring of venting burning gas.

I watched in awestruck wonder as a continent was calved before my eyes. The massive ring of fire continued to grow and its sooty plume mushroomed up, and then folded over to spread toward the tropical continents. Suddenly it dawned on me. "That's a giant version of the same phenomenon I saw on the test drive this morning! Venting of methane hydrate will always break glaciation cycles. Earth has a built in freeze protection system. Every time it glaciates, it will release buried methane. What a fantastic design."

"Trust me, I wouldn't leave you in the freezer." Ralph shared my elated mood. "A few moments ago, you wanted some greenhouse gasses to reappear and a reduction in reflectivity to warm this place up. What did I deliver?"

I looked over the continent sized ring of fire and imagined a similar situation at the North Pole. The surrounding fractured ice was blackening from the sooty plume. The direct heating from megatons of burning methane was a fast way to start melting ice, but the increase of heat retention from a slight change in atmospheric composition would add a lot more heat in the long run. Here was a source of instant heat of combustion, a decrease in reflectivity, and an increase in atmospheric methane and carbon dioxide all at once. "You over-delivered. Thank you."

"Don't thank me, I'm just the tour guide. Let's take a fast trip through some more thermal cycles.

Often in science, when something is understood, it is unbelievably simple. I had struggled for years with the extraterrestrial reasons given for

Earth's warming up from a glaciated condition. Small changes in heat input would not thaw out Earth as fast as scientists now know it thaws after an ice age. I had my answer. It was simple and beautiful. It was going to be fun to watch it happen again.

Chapter 16

THE ICE MELTED. AFTER THE tropical land mass shed its load of ice and snow, the crust slowly rebounded. Tremendous forces generated from stretching of the rebounding crust created cracks. Pieces floating on the lubricating molten lava below began to drift apart. Once again continents roamed the globe.

Earth cycled between two more freezing and thawing cycles as we sped through time. It was getting repetitious and other than repositioning of continents between cycles, the results were the same ... a slow freezing followed by a fiery transition and a rapid thaw. This seemed to be very similar to the pattern of the recent ice ages that scientists could study through examination of ice cores. I wished that I could see the graphic that I had left in the folder in the condo. No sooner had I thought it than Ralph nudged me ... he had it up on the display.

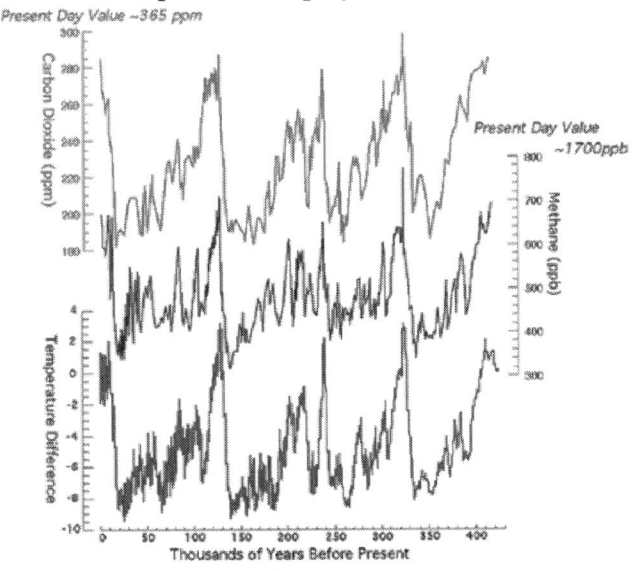

"This is just about what we are seeing now!" I nearly squealed with excitement. "Earth cools down slowly, freezes deeply, then rebounds from its icy state when the pressure is reduced on sub-oceanic methane hydrate causing it to decompose and flood the atmosphere with methane."

"Except with the weaker sun at this time in history, the freezing takes place faster," Ralph said, "and with little oxygen in the atmosphere methane lasts longer."

"And makes for a shorter heating cycle too," I interrupted with excitement. What a beautiful and simply elegant way for Earth to avoid freezing at the time of a weak sun.

We sped through several freeze-thaw cycles over hundreds of millions of years. Some were minor icings that recovered quickly and others more extensive. All were broken by the release of methane hydrate and none of them warmed enough to significantly heat the ocean, or change Earth's ice age cycling process. Continents roamed about, volcanic activity continued, and landmasses increased in altitude. Life thrived in the ocean, but did not spread onto the land.

"What's wrong?" I asked. "Earth appears to be trying for something it cannot quite achieve. How long will this go on? Do we need a major event to break these never ending cycles?"

"Patience, Earth still has needs and this process is supplying them," Ralph implored. "When the time is right we'll get the "Big One.""

"What needs does Earth have?"

"Oxygen for one. It is increasing, but we need more to support a protective layer of stratospheric ozone and allow the start of terrestrial life. Your Earth also needs fertile soils that this volcanic action and glacial grinding is producing. And … if we stopped freezing right now, where would the carbon to make trees, flowers and plants come from?"

Wow! What an eye opener. Somehow a massive carbon source would have to supply terrestrial earth with carbon to make all the plants, animals, and coal that would be created in the next hundreds of millions of years. Where would it come from?

"Let's go see it happen." Ralph grabbed the controls.

Chapter 17

WE CRUISED THE WORLD TO see where the continents had drifted to during the past eons of freezing and thawing cycles. The continent of Antarctica had risen to a height that held snow through all the warming and cooling cycles. In the north, there was a frozen polar ocean surrounded by continents and islands.

We slowed our speed through time. "We are entering another cooling cycle." There was an edge of excitement in his voice.

When we passed over Antarctica, it was completely snow covered. The Arctic Ocean was frozen over and nearby landmasses had accumulated year around snow. We cruised an equatorial orbit watching the scenery go from ocean to land, to ocean, to land. I could not relate to Ralph's growing excitement.

"We got it right this time!" He was really excited. It was as though this was his first time here. I didn't get the significance of whatever we were looking at. It looked like the start of any other cycle to me.

He didn't even force me through a tough thought process, He babbled, "This continental arrangement supplies strong ocean currents to carry warm equatorial water clear across the northern polar ocean and gives a strong circulation around the Antarctic continent. These warm waters will supply the freezing polar air with water vapor to drop huge snow deposits on landmasses. The ice cap on land will continue to build up while the oceans stay open to supply water vapor for more snow and ice. The ice caps will extend well below forty degrees latitude before ice forms on the polar seas."

I guess this was a setup for the "Big One."

He ran us through time. The world below responded as fast as he talked, doing exactly what he was saying. He continued. "Meanwhile, the equatorial oceans are still warm and productive. They support biological processes and keep sequestering the carbon dioxide from volcanoes and methane venting from the oceans."

Volcanic activity over fields of snow provided some spectacular scenery. The cold earth seemed to be approaching a balance of atmospheric gasses, but it was still cooling. How far would it go this time? With Earth's reflectivity increasing, the temperature would drop until the oceans productivity was impaired, making it unable to keep up with the supply of greenhouse gasses.

Snow kept creeping from the poles toward the equator. If Florida existed, it would have been snow covered. Miles thick ice sheets moved to the south where they stagnated as the oceans near them froze and blocked off their snow supply.

On a trip north to check out the frozen Arctic Ocean we found the ocean level had dropped much further than before, but there was no indication of methane venting, just piles of cracked ice. I went down and checked the analytics to find the methane levels elevated, but not significantly.

This was disturbing. Arctic methane venting had led previous warming cycles. What now? "What happened to the methane?"

"The last venting removed most of the available methane hydrate from the Arctic waters." Ralph sounded almost gleeful. "There isn't enough methane hydrate in the Arctic to make a difference this time."

"So we will get colder and colder." It was beginning to feel like the "Big One" was going to be a big freeze.

"No te preoupes amigo." Ralph laughed. "There's more methane hydrate available in other oceans."

"Earth will have to go deeper to get methane hydrate from areas that had not been disturbed before," I guessed.

"And?" I could feel the smile in his voice with this coaching "and."

"It is probably like mining a rich new lode," I ventured. "Methane hydrate deposits that have survived many freeze-thaw cycles, will become unstable as ocean levels decrease, and ocean floors rise, to vent methane into the atmosphere at an overwhelming rate. Ice Cube Earth is dead meat!"

"It may be more spectacular than you think, let's take a look. I believe I remember where the real action started."

He took the controls and spotted us at about ten degrees north and two hundred thousand feet. The weather was clear. To the north, where warm open water supplied vapor to the ice covered edge, a lightning storm covered the horizon. The calm ocean was an expanse of undulating blue-black water.

An area of whitening half way to the storm line caught my eye. It looked like it was boiling and rapidly growing. "Methane?"

"By the megaton."

I surveyed the situation and winced.

"Scared?"

"Yes." I shuddered to think of what would happen when lightening and methane got together. "Trust me, I've been here before. You are about to witness the biggest fireworks show on Earth."

I backed into the center of the craft and wrapped myself into a shivering ball. Were we safe ... even in this high tech spacecraft?

"Relax," came the comforting voice. "I want you to have a good look."

I was foolish to worry. He had been here before and survived. I stepped forward for a closer look.

The ocean below was foaming, lifting the blue-black quiet waters to raise them up as froth, flow over the surface, and grow the area of foam. The white frothy disk radiated from its starting point towards the horizon and approached the waters directly below us. It was also moving towards the lightning.

I felt us drift south and gain altitude. Evidently Ralph wanted to be near the impending blast, but not in it. I was comforted by his precaution.
I watched the lightning, wondering when it was going to strike the flammable mixture of gasses that were getting ever so close. Close, I thought, I'm looking at thousands of square miles of flammable gas that must reach miles into the sky. This was to be the largest non-nuclear explosion ever.

Suddenly one of the lightning strikes turned orange instead of dissipating. The orange color swept over my entire field of vision. It was like an action movie screen suddenly filling with ugly oranges, reds and browns. This was not a movie screen; it was a state sized area. I pulled back from the window.

"Watch," Ralph encouraged. "The ship will rattle, but we are safe." I was glad for the warning. The initial flash was spectacular, but it was just the preamble. It must have served to compress and warm the gases below that suddenly exploded in a hot white flash. The ship shook violently.
"Wow, I wonder how hard this shock wave hit Earth?"

"Good thinking. Look closely." The spacecraft rushed in to get a view of the center of the blast area.

Once I stopped shaking, I looked down to see dry land. It was like

Moses had parted the waters, but these waters were hundreds of feet deep. A mountainous wall of water was radiating out in all directions from the blast center. For a fraction of a second, I saw the dry ocean bottom, then red hot lava below as the thin crust of the young earth was split. I looked up to see tsunami waves racing away. Within minutes waters collapsed back into the giant hole to send steam, fire and ash skyward in a slowly developing mushroom cloud. The fires dissipated as the cloud topped out well into the stratosphere. The tsunami waves were heading for the far horizon.

"This is going to wreak havoc all around the world," I muttered.

"What kind of havoc?" Ralph wanted specifics.

"Those waves are going to scour the coastal shelves and stir up methane hydrate. There will be a lot of methane going into the atmosphere. It's going to get hot and stay hot."

"You wanted to warm this place up and get rid of this ice, didn't you?"

"Somehow I thought that this would be a gentle process." I was numb from watching the explosion and the awesome mushroom cloud. I wondered what those giant tsunami waves were going to do to the Arctic ice.

"I'd like to see the polar region." I no sooner uttered the words than we were hovering over the expansive ice covered Arctic Ocean just before the tsunami rushed between the northern reaches of two continents. The ice surface spackled into millions of pieces as the water level in the broad strait dropped in front of the approaching tsunami. The imposing wave front threw a shower of ice into the air as it swept across the Arctic Ocean. We went to a higher altitude as the line of exploding ice faded into the horizon.

"We need to speed this up a bit," Ralph said and I watched the indicator go to twenty times normal speed.

When the tsunami exited the strait, it fanned out and diminished slightly, still throwing ice high into the air at its front. Subsequent waves bouncing off landmasses crossed the Arctic Ocean to grind what was a formidable ice sheet into piles of icy rubble. It was an awesome display. I was transfixed by the power and resulting beauty as the Arctic Ocean became filled with blue and green-hued piles of shattered ice.

"There's more to see." Ralph maintained our altitude and pulled us away and to the south, just off the shore of a major continent.

The wave had rushed hundreds of miles onto the land. I was mesmerized as I continued to focus on the amazing volume of water carried to the inland horizon.

"Look at the coastal oceans."

I took his advice as the wave receded and looked offshore. The ocean continued frothing well after the violence of the retreating wave subsided.

Methane hydrate? It seemed to be a more vigorous frothing than I would have expected from a short-term pressure reduction resulting from a momentary reduction in water depth. There had to be something else going on. I looked closely. Occasionally, I noticed white solids the size of basketballs at the surface.

Did the tsunami rile the ocean floor enough to release large pieces of methane hydrate and bring them up to the surface? Then I remembered that methane hydrate was extremely light. It was only thirty percent as dense as water and any broken pieces would rapidly float to the surface, taking some of their sediment matrix with them.

"We got help from an underwater landslide, just like in Cussler's *Fire Ice*." Ralph increased our altitude. "Let's take a look around the globe."

We gained more altitude and raced in an orbit that would hit fifty degrees north and south. "I'll do a loop, then one each at sixty, seventy and eighty degrees," he said. "That will just about cover it."

Great. I wanted to see if all ice was as busted up like what I'd seen in the Arctic and just how much of the ocean was venting methane hydrate.

The four orbits afforded a good view of most of Earth's surface. Only in bays or deep fjords was the ice intact. Methane bubbling was taking place on less than five percent of the ocean's surface. What an amazing change in a short period of time. What would happen now?

The tremendous amount of energy released by burning methane raised the temperature immediately, but the amount of soot and debris in the atmosphere blocked out the sun for years. Energy was being blocked from getting to Earth's surface, but the sooty atmosphere itself was warming and once it cleared, the soot-covered snow was a rapid heat absorber. The still venting and burning methane continued to supply heat. Was most of the methane scoured from the ocean?

The analytics screen came up to show that methane and carbon dioxide had risen, but nowhere near to what I thought they would from the dramatic venting. The oxygen levels took an almost imperceptible dip.

"Quite a dramatic display of energy." Ralph's voice was soothing, sounding as though he was satisfied with the outcome. "But the ocean didn't give us all of its carbon."

Was this not the "Big One?" Would we refreeze? What would happen?

"What happens next will happen slowly, we will speed it up from here."

The speed dial rose through twenty-five years-per-minute. Piles of broken-up ice were stacked up on shorelines as storms generated by warm open oceans moved northward. Ice capped landmasses shriveled away under the rapidly increasing temperature and rainfall.

I watched the time meter spin up to an even faster rate. Ralph, the tour guide, was in his glory. He had obviously been here before and he knew what was needed to break Earth's freeze-thaw cycling, but his joy in seeing it again was almost like that of a child on Christmas morning. Whatever his function, he obviously loved his job. As "Earth Keeper," was he more than an observer and guide? Could he influence outcomes?

"My capabilities of influencing anything are nearly spent, but Earth has been by far my most successful project. You are going to love seeing how it comes together."

Interesting, I thought. He is more than a guide. I was going to probe further, but Ralph interrupted. "It's time for you to focus on the oceans." This was delivered as a crisp directive to stop meddling and get to work.

Much of the ocean areas had been iced over and quite nonproductive for centuries. Was the ocean dead?

"Worried about a dead ocean?"

I thought a moment. It had been dead before and there was a lot of life down there ready to thaw out and get to work. "No," I said emphatically.

"Good answer. Now watch the changes."

With the ice cover gone, ocean currents reestablished themselves and began carrying warm water north. How warm was it? A thermal map indicated that tropical water temperatures were at 95 degrees Fahrenheit and cooled to 55 by the time they reached the Arctic Circle. Not a lot different than home.

"Let's get back up to speed. What's going to happen now that the ocean currents are established?"

"We will restabilize," I said, thinking that we were due for a balance similar to what I had left in Cancun.

"Really now?" His tone indicated that I'd better start thinking and watching. We probably were not headed toward stability, or if so, it would be a new stability.

He increased the speed through time and we went north to get a look at the termination of the ocean currents. "Take a look at the temperature of the cold end of the current." I watched the temperature map as Ralph sped through time. We were not stabilizing. The temperature rose to sixty, sixty-five, then inched toward seventy degrees Fahrenheit.

"Any concern?" he asked.

"The warmer water is releasing carbon dioxide," I said, "But we should be able to handle that with increased biological activity." I saw no reason for special concern.

I sensed that Ralph was about to respond with another "Really now" when I thought of methane hydrate. I really wished that I had a copy of its phase diagram to study. With the thought barely finished, the diagram appeared on the display.

The phase diagram indicated that methane hydrate would decompose at any ocean depth if the temperature were above seventeen degrees Celsius. I converted that in my head to sixty-three degrees Fahrenheit. As the oceans warmed, methane hydrate at the ocean floor, and even deeply buried methane hydrate would become unstable and dissociate. Free methane would escape to the surface. Or, if a sediment layer contained the released methane, it would build up and increase in pressure until a catastrophic release broke through.

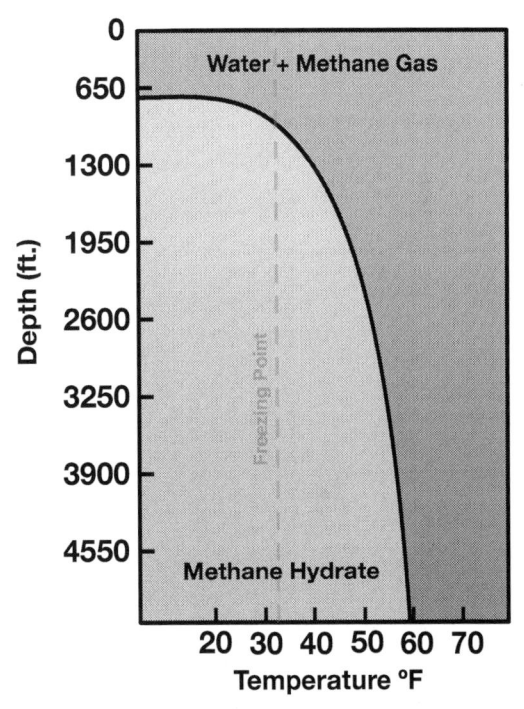

I scanned the area below and saw bubbling, or fire, define areas where a steady flow of methane was venting from the ocean. Ralph sped up the time. Bubbling or fiery waters mapped out areas of the oceans warm enough to decompose methane hydrate. There was going to be a lot of energy transfer from stored methane hydrate to the atmosphere. We may be heading for a stable period, but it would be a warmer stable period than I had ever seen.

The atmosphere was oxidizing methane, an extremely powerful greenhouse gas, to carbon dioxide, but the amount of methane venting from the ocean was overwhelming the atmosphere's capability of keeping methane concentration in check. Storms intensified with increasing ocean temperature, but what was happening in the oceans? Would oceanic life respond and bring the runaway greenhouse gasses into balance?

Biological activity had slowed when the oceans cooled, but there would not have been a complete die off. Would the warm up be like taking an active culture out of a refrigerator? Could there be a resurgence of oceanic life?

"Be ready for a surprise." My thoughts were read again and we slowed down to real time, dropped our elevation, and cruised near shore islands and bays.

I took the controls for a closer look at what appeared to be an accumulation of color at the shoreline of a deep protective cove.

"Does that look familiar?" Ralph asked.

"Sure does, just like the last time. Biological activity is coming to the rescue and will remove energy from the system and cool this place down."

"Does it look exactly like the last time?"

From his tone, this was another clue. I looked critically. The fibrous ebb and flow of green strings of material looked the same. Where was the difference? The green seemed to be flowing onto the shoreline. I did not remember that before. A closer look showed a green mat establishing a beachhead. Was it living outside the ocean environment?

I looked behind the dune of a broad beach exposed to the open ocean. There were scattered pieces of green being tossed around by the surf on the ocean side. Did any make it over the dune and establish in an area not exposed to waves? The deepest pockets, sheltered caves, places that would be the wettest, were tinged in green. Were they nurtured by ocean spray? Could storm tossed life survive far from the shore?

I went inland and searched pockets of low land and saw, or imagined, the establishment of biological growth. We saw life established in the highlands, in the plains, and in fresh water lakes. The ocean was going to get some help in removing the carbon it was venting to the atmosphere. That should slow the warming a bit, I thought, and wondered what the next surprise would be.

"We will have to get back to the ocean to see what's next." Ralph took over and brought us back to the topical oceans and sped through time. He nodded toward the temperature map. The ocean was approaching a hundred degrees. What was he trying to tell me?

When the ocean cooled, it took in a lot of carbon dioxide from the atmosphere, became more acidic, and dissolved calcium carbonate. Now, were we about to see a reversal? As the ocean warmed, the water softening reaction of converting dissolved calcium and magnesium bicarbonates to carbonates was about to take place on a grand scale. The warm areas of the oceans effervesced with released carbon dioxide and turned chalky as millions of tons of carbonates filled the waters.

$$Ca(HCO_3)_2 \longrightarrow CO_2 + CaCO_3 + H_2O$$

(Calcium Bicarbonate) (Carbon Dioxide) (Calcium Carbonate) (Water)
(Dissolved) (Gas) (Solid) (Liquid)

Calcium Carbonate would eventually settle and turn into sedimentary rock. "Is this the start of pre Cambrian cap rock formation?" I stated one of the few nodes on the geological timeline that I remembered.

"Good guess. Now what happens?"

I scanned the diagnostics. Atmospheric carbon dioxide shot up as it vented from the ocean, and the temperature was still edging up. Oxygen levels were rising. The warm ocean was regaining its productivity. Were we coming to a stable point? So many things were different during this recovery from a freeze up that I had no idea what was happening. I remained silent.

"We need to talk," Ralph said.

The Long Stability

Chapter 18

I REALLY LOOKED FORWARD TO listening to Ralph. How much would he reveal? How many different life forms were present in the various green globs I had viewed? Did life on land mean that a protective layer of stratospheric ozone had developed? Where did this life come from? How much carbon remained in the cold deep ocean as methane hydrate, or oil and methane formations capable of supplying methane hydrate? Were we in a period of reheating and cooling, or would we reach a new stability?

"Hey … wait a minute," Ralph chimed in. "When I say that we should talk, I really mean that you should take a time out to think while I listen."

I should have known that free information was not on the agenda. I waited for a clue.

"Think about what has changed since that last time we had an Earth this warm. Start with the atmosphere itself."

I thought a moment and was surprised at what popped into mind. First of all, it is cleaner. Bombardments from large asteroids were still significant events, but becoming rare in frequency with short-lived and localized effects.

The atmospheric composition was also very different. Sulfur in all forms, free hydrogen, and carbon monoxide were virtually gone. Oxygen levels were still increasing and the last time there was none. The atmosphere was really different.

"Good job, now how about solid Earth? Has it changed much?"
I thought back to the days of the start of the oceans and the struggling landforms solidifying on the molten surface. "What drastic changes! We have cold continents, mountain ranges, fertile plains, shallow inland seas and lakes, and well-developed tectonic plates to grow mountains and support

continuing volcanic activity to fill valleys with nutrient rich soils. Earth is developing a vast diversity of landforms and soil types." I paused, thinking that was about it. The ship was silent.

"Last time, the atmosphere had to perform a major change on the Earth that is not necessary this time," he finally clued.

What could that be? Wow, it was major. "We had to do a lot of oxidation of minerals the last time. That's behind us. We can really create an oxygen atmosphere in a hurry this time." I found myself feeling like a partner in the creation of the Earth's attributes.

"How about the ocean? Is it different this time, partner?" He chuckled.

"The ocean currents are well formed around the continents and heat is rapidly transported to the poles. The water is still warming up, but it shouldn't go much warmer now that the atmosphere has stabilized and the ocean is beginning to remove carbon dioxide." I was rambling around the fringes, talking about significant things while having the feeling that I was missing the main event. "This time we are starting with life present," I babbled. "Probably a variety of life forms." We had seen life at the surface and I could not imagine that making Earth's surface into a snowball would affect life at vents or hot spots on the ocean's floor. There had been millions of years for life to form and evolve in response to changing conditions. I thought of the rapid trauma that this life was subjected to as the warming began. "Whatever existed in the cold sure got a shock when the methane vented and locally purged a lot of oxygen from the water. The warming caused venting of carbon dioxide, and the resulting change in acidity was a major shock to any and all of the primitive life forms present."

"Would that be disaster?"

"More likely that the stress resulted in adaptation and morphing into stronger and more resistant forms." I could not contain myself. "This place is ready to explode with life!"

"That's why I picked it," he said with a hint of pride. "Everything went well in the early period of Earth." His tone saddened.

He must have thought something turned bad at some time. "From an Earth Keeper's perspective, did it go bad? What and when?" I knew that was none of my business, but thought I'd probe.

That thought brought a chuckle. "Your business is far from complete, and your night is waning. If we get your business done, we can talk more about mine. The next eras are very exciting, but well studied from their fossil records. For your purposes we can whiz through them. Ready?"

We cruised ocean currents starting out at near one hundred degrees Fahrenheit at the equator. The Arctic bound currents were visible torrents that spun off tremendous storms on their way north. At the northern stretches, the currents cooled to seventy-five degrees.

"It looks like most any life that could have been generated so far will find a place to survive this heat up," I ventured, looking for any clues he would give.

"Everything survived and will rebound from the snowball in an explosion of morphs and species."

Chapter 19

"WELCOME TO THE CAMBRIAN EXPLOSION of life." Ralph was in his glory as we swooped around the world and poked into interesting crevices and bays. Even the brightest days provided only a twilight-like glow through the thick and cloudy atmosphere with spotty bits of sunshine, but life seemed to be everywhere.

"Is this hothouse temperature going to be stable?" I could not believe it, but most of the atmospheric analytics were stabilizing.

Methane levels had jumped up from the initial release, then decreased and stabilized. Carbon dioxide increased with the initial venting and burning, then edged up a little more when the oceans began releasing carbon dioxide from dissolved calcium carbonate. Plant growth on the land surface was utilizing the increased supply of carbon dioxide and creating more oxygen, but atmospheric oxidation of methane to carbon dioxide was keeping oxygen levels from a rapid increase.

Temperature increased again as Earth transferred tremendous amounts of energy and greenhouse gasses from its sub-oceanic storehouse into the atmosphere. The oceans slowly and steadily warmed. In looking at the chaotic recent history, it appeared that everything but oxygen was stabilizing. Would the temperature increase and then remain stable?

"More stable than any period so far," Ralph responded.

"Why?"

"You tell me." He took us to altitude and toggled the view port filters to allow observing the ocean below. "Watch the currents."

I looked at a warm current streaming up from the equator toward the pole. In the tropical regions, the current was bubbling. "Is carbon dioxide still venting from the heating of bicarbonate filled waters?"

"Yes. What we initially observed was only the beginning of the carbonate deposition. A lot of ocean waters remained cool and bicarbonate rich."

We followed the warm stream north to where it cooled in the polar region, dipped to the ocean floor, and returned south. A long thin line of white bubbles marked its return path as the 'cold' returning waters warmed enough to release methane hydrate. "Amazing!" I shouted looking over the ocean's continuing mechanism for releasing carbon dioxide and methane to the atmosphere. "This will last for a long time."

"I'll fast forward you." He went through a few millennia and it was clear that the current was moving to the east … the continent was moving too. "For the next few hundreds of thousands of years, the recycling of the ocean floor through volcanic action, and the sweeping of warm ocean currents will give the atmosphere a steady supply of carbon. We will have a lot of variability, but we will not cook life off the planet and it will be a long time before the next major freeze."

"And when it runs out?"

"No te preocupes … enjoy the ride."

We sped through seventy million years of steamy weather. Atmospheric oxygen continued increasing as plant life on land spread and turned carbon dioxide into oxygen. He swooped down to the surface of a very large bay in a continent at about forty-five degrees north.

"We have time for a short side trip. I'd like to show you another reason I held out for this later model."

My stomach rose to my throat as he dove into the water without as much as slowing down. I braced for a jolt, but there was none. We dove through green tinted water with a healthy suspension of algae and descended rapidly. There must have been thousands of different species, most in schools ranging from a half a dozen to thousands. At the ocean floor, corals grew in massive gardens interspersed with a variety of shapes, sizes and colors of sponges. I wished I had a frame of reference for the current understanding of this era.

"Don't fret. This era is really quite well documented. Your scientists have identified about 900 life forms."

"That's a start. We've looked at a sliver of ocean and seen thousands. There are probably much different colonies at the equator and the poles."

Ralph stayed under water and we cruised past an abundance of life as we sped through time. It would have been fun to stay there for a long time, but I was getting the point. Life in the ocean had exploded in a variety of forms.

"There has been amazing progress since the last time the ocean assimilated a carbon filled atmosphere," I shouted. "Back then it was only a few crude organisms capable of one necessary job, sequestering carbon in the ocean. Now look at the variety of plants and animals."

"You sure are catching on, but you haven't seen anything yet." We burst through the surface and headed into rapid time.

Chapter 20

WE WERE ENTERING AN EXCITING era and I looked forward to seeing the development of complex plant, animal and human life on land.

"Remember your mission," Ralph said. "Earth has done a lot to bring us this far. What has to happen to get us to life as you enjoy it?"

I had to stop thinking of the development of life and stick with the environment that supported life.

"Good thoughts. Now where are we in the development of human nurturing resources, and where are we going?"

"We've certainly transformed rocks into soils through volcanic and glacial actions. We've converted carbon oxides into organic compounds that are now developing into complex oceanic life and will soon generate an amazing variety of terrestrial life. Right now, we do not have enough oxygen for man, and it is too hot."

"There is also a steady supply of carbon dioxide and methane coming from the ocean," he reminded me. "Is that a problem or an opportunity?"

"If it cannot be transformed into limestone, plant matter, and coal or otherwise sequestered on or under the land or oceans, it will accumulate in the atmosphere and eventually drive life from an overheated Earth. If it can be handled, it will supply Earth with a variety of organic material for the life I now enjoy."

"Exactly. Now I have to warn you. We've been having too much fun and we are running short on Cancun night. We'll have to go through this era with strict adherence to your agenda."

That was a sad message. I had hoped to see the rise and fall of the dinosaurs and the development of modern animals, to say nothing about the rise of man ourselves.

"Your science has the periods from here to the future well documented.

There is no need to spend time rehashing known ground."

I knew that from my reading. I also knew that there were many conflicting opinions on some of the major events. I'd sure like to get the straight story.

"Just what I don't need." His voice was edgy. "It would not do either of us any good for you to go back and straighten out the experts."

"Do we have time for an overview? I would hate to miss the period entirely."

We went to a two hundred mile altitude and sped through time on a pole-to-pole orbit. As Earth warmed, it became steamy with water vapor and heavy with cloud cover. Most of Earth's surface was blanketed in rain and storm activity. Our cloud and rain filtering view port was engaged much of the time.

As we sped through time, it was apparent that volcanic action continued supplying new land surfaces, tectonic plates bumped into each other to raise mountains and the heavy rainfall wore mountains down. Occasionally, ice accumulated on continents in the far southern hemisphere to lower global water levels. This caused the draining of inland seas on major northern continents. When the ice melted, global water levels rose to cover more shoreline and fill inland seas.

Mega volcanoes filled the sky with sun blocking particles to cool Earth off, but that also slowed the removal of carbon dioxide from the atmosphere and Earth's heat rebounded. Other periods of increasing greenhouse gasses and heating resulted when their supply outstripped Earth's capability of sequestering them. Continent-sized fires, or forest-destroying plagues resulted in warming. Desert winds sending nutrient laden dust into the oceans touched off growth rates that led to cooling. It was a wild ride, but there was always a bountiful supply of ever changing life on much of Earth's surface and I'm sure that extended into the oceans.

With time it was becoming apparent that even though this was a stable period without dips into major ice ages or overheating to eradicate developing terrestrial life, there were major variations with time. As earth greened and larger and more complex plant forms evolved, the tropical areas were the first to develop. With increasing warmth the polar regions flourished as portions of the tropics turned into deserts.

I played with the analytics history for a while, looking at the gas composition and temperature. In all cases, the temperature followed the concentrations of carbon dioxide and methane, with the greatest sensitivity being to methane.

"And you expected?" Ralph asked.

I really expected that temperature would follow the greenhouse gas concentration, but all the controversy in my Cancun Earth had my faith wavering. It was good to see that history went according to expected science … blanket Earth with heat trapping gasses and it gets warmer.

"Take a look at the effect of temperature on environment," Ralph said.

We hovered over a tropical area. The variation in environment went from desert to savannah, to jungle, to a lake, and then backs to a desert. In checking the polar regions, they ranged from near tropical climate to chilly temperatures that limited or nearly destroyed their tropical footholds. In the mid latitudes, growth conditions prevailed through all temperature swings, as growth types varied with climate. Growth on growth activity buried carbon under Earth's surface.

At the speed we were traveling, it was apparent that any life form had to be capable of geographically moving with or adapting to changing climate. "We're staying in somewhat of a balance," I said, "but plant and animal species sure have to possess nimble migration or adaptation capabilities."

"The survivors anyway. Are they adapting to anything other than climate and geography?"

Survival is also dependent upon food source availability, predators, disease, and pests, I thought, but we cannot see that from here. I checked the atmospheric analytics. There certainly was a cycling of methane and carbon dioxide with time, but oxygen was steadily increasing. An increasing oxygen environment was a major change that any life would have to tolerate, evolve and adapt to, or die off.

I thought back to a book, *Under a Green Sky,* which I had recently read. The author had spent a career studying major extinctions and concluded that runaway climate had caused all but the one extinction at sixty-five-million-years-ago that was definitely asteroid induced. We were traveling fast. Were we seeing these extinctions?

I didn't get the thought past Ralph and he immediately asked: "What do you think? Are the changes we are seeing rapid enough to leave a large fraction of life behind?"

"Certainly seems like vast areas are purged of one habitat and transformed to another fast enough to affect all but the most adaptive."

"What was the major killer in his explanation?"

I had to think for a moment. "He had Earth warming and the

temperature difference between the tropics and poles lessening. This sent less energy poleward and disrupted ocean currents. Tropical oceans warmed and surface water concentrated in salt. Oceans stagnated, lost their oxygen and produced hydrogen sulfide. The dense warm tropical surface waters sunk to drive ocean currents and cause the venting of hydrogen sulfide. Hydrogen sulfide killed life directly and entered the stratosphere to destroy the ozone layer." I thought over this reasoning. "Did that really happen?" I asked. "It sounds like overkill. Like too much destruction."

"We are not seeing everything on this trip and I'm not saying it didn't happen. I can say that hydrogen sulfide was not the major killer in most partial extinctions. Remember, we are not here to prove of disprove a bunch of current theories. How did *Under a Green Sky* cope with methane hydrate?"

I never thought of that when I was reading the book. All of a sudden the effect of sinking warm tropical currents sunk in. "The sinking warm currents would decompose a tremendous amount of methane hydrate! That was never mentioned."

"Your current scientific process is slow to catch onto anything slightly removed from very deep and narrow fields of study." His voice turned almost mournful. "That is probably one of our major shortcomings."

What did he mean by that? How could he influence the field of scientific study? The closer I observed academia, the more shackled it seemed to me, but I thought that was my prejudice gained from a long industrial career. I'd like to learn more about this phase of the Earth Keeper.

"Not now," Ralph said, "We've got ground to cover."

A diversity of survivors continued to populate the fertile planet. I checked the clock and found that we were closing in on sixty-five-million years ago, the time of the extinction of the dinosaurs and the depositing of iridium around the globe from some dramatic event. There was controversy as to where this event took place, and I really wanted to see this turning point of history unfold.

Suddenly I felt light-headed. I wasn't fearful. I felt my pulse and it was strong. What was happening?

Chapter 21

"Wнат was that?" I mumbled from a deep mental fog while trying to rise from my chair.

"Sorry, we are running late and I accelerated you through time at a humanly intolerable rate. My fault."

I'll bet, I thought. He probably didn't want me to see the great extinction. I looked at the time and we were five million years past it.

"We're just in time to see Earth really warm up. I thought you'd appreciate a close look."

Warm up, I thought. How could it get much warmer?

We zoomed over the broad strait between the Indian and Eurasian continents, stayed stationary, and flew through time. The continental crash was the most spectacular I had ever seen as the dividing ocean released methane for hundreds of miles and slowly disappeared. Even more impressive was the action further out in the ocean. As the ocean floor buckled under the stress of the collision, great eruptions of methane foamed out of the ocean, and ignited to light up the sky for decades.

Methane and carbon dioxide levels shot up. "This is really going to heat up the ocean," I said pointing to the atmospheric analytics.

"Think about the positive feedbacks. Over the next few million years the deep ocean temperature will rise five degrees Celsius." (Nine degrees Fahrenheit)

I remembered a graphic of this era that I had in my folder and wished I could look at it now to see if it reflected what was going on in front of my eyes. Ralph supplied it on the display screen.

Global Deep Ocean Temperature

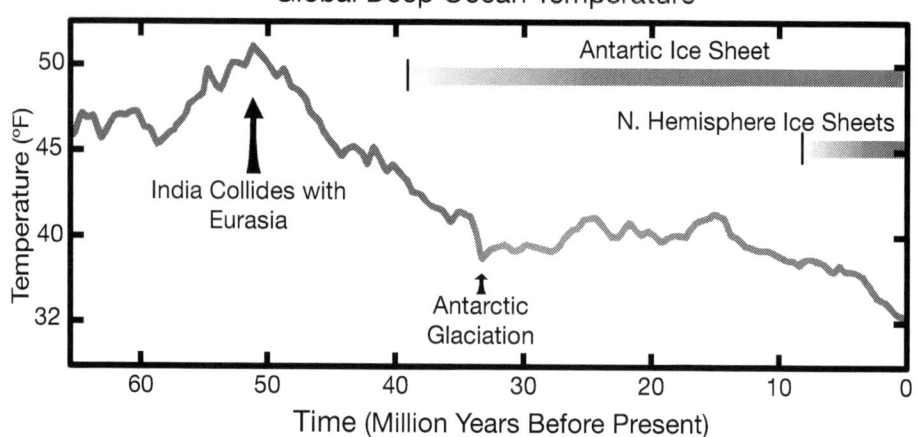

(Chart: Temperature (°F) vs Time (Million Years Before Present). Features labeled: "India Collides with Eurasia" peak around 50 million years ago near 50°F; "Antarctic Glaciation" around 33 million years ago; "Antartic Ice Sheet" bar; "N. Hemisphere Ice Sheets" bar.)

With ocean temperatures rising and much of the ocean full of methane hydrate on the verge of stability, any heating would bring forth more methane causing more heating and on and on. What a positive feedback this was. Absolute runaway.

"There is another major positive feedback about to take place." Now I caught on. "For the deep ocean to rise five degrees, a lot of volume above it will reach temperatures where calcium bicarbonate is no longer stable."

"The good old reliable water softening reaction," he said. "Let's go see it."

We took a series of forty-five degree orbits to see the variety of oceans between forty-five degrees and the equator. Large chalky rivers appeared to be running through the open ocean as water temperatures reached the point where calcium bicarbonate released carbon dioxide and precipitated out as calcium carbonate. It would continue to get warmer.

"Until when?" Ralph asked.

"Until we run out of available methane hydrate and the oceans purge themselves of carbon dioxide," I speculated.

"Very true. Anywhere special you'd like to visit?"

"Axel Heiberg Island! I'd love to see that at its warmest."

I had seen preserved remnants of large tree trunks from this era on a trip to the Canadian High Arctic and wondered what it was like when forming. We were there in a heartbeat.

It was spectacular. All the northern islands were covered with gigantic sequoia and spruce. Dolphins and whales cavorted in the surrounding waters. Large amphibians crawled around the beaches. Insects the size of swans were buzzing through the air and being perused by matching sized birds of prey. It was an absolute tropical paradise.

What a difference from the August snowstorms over desolate landscapes I had experienced. The largest live tree I had seen was an eight-inch-high willow the diameter of my thumb. If it got this warm at eighty degrees north, I wondered what the tropics were like.

Ralph sped us south to find desertification spreading well into subtropical areas. The temperature stabilized for a while, and then began to slowly cool as we sped through time.

We were approaching the five million year mark. I needed another reality check. Seeing the Arctic forest where I had visited the preserved trunks in 2000 was very helpful, but I needed more carbon deposit reference to tie this trip into reality.

I took the controls and hovered above the Grand Canyon. Many years ago I had been here in the winter. Gloria and I visited her parents in Phoenix and her father and I went north to drive along the canyon wall and marvel at its width, depth, and beauty. Its snow-dusted rim accentuated the rugged dark rock formations. We took a ride in a light airplane to get some spectacular views, a treat that is now banned as disruptive to nature.

I wanted more and later I rafted the Colorado River and was astounded by the open wall of time that unfolded as we dropped a mile in altitude over a week's time. I had carried a geological guidebook and for a while remembered the names of alternating layer of ocean formed corals, and sedimentary sandstones. I also remembered that the river bottomed out at 1.7 billion years in a hard schist ... whatever that was. What was on top? How old was its youngest formation? I had no recall.

I looked at the canyon with renewed awe. I was not really watching a spot of land over a long period of time on this trip. I still had little concept of how various areas changed with time, from ocean to sand dune and back to ocean.

"I didn't take you to watch the continental plates move and the rising and falling of the crust with time. That's a minor part of your needs. If this cut were made in another part of the world, you'd be looking at the old forest areas now as coal seams. You would see oil shale and tars formed from the heating and compression of organic matter. Why did you come here?"

"I'm sorry, but I need to get this into perspective. How old is that massive top layer of limestone?"

"No te preocupes, that's 250,000,000 years young." Ralph laughed. "It is time for you to recap and think.

The Modern Cycles

Chapter 22

I DIDN'T WANT TO WASTE the time it took to see the Grand Canyon, but it helped. The top formation was 250,000,000 years old and here I was only a mere 5,000,000 years into the past. There was a lot of development of the Earth to complete, including the recent ice ages that were initially of most interest to me. In seeing the canyon, I realized how little of Earth's creation I had seen on this trip, but I did get a close up of what I really wanted ... an understanding of how the atmosphere changed and carbon transferred between the atmosphere, oceans and land in cycles that made sense. We needed to get through the last five million years before Cancun dawn. I hoped that we could make it.

"No te preocupes, amigo," Ralph assured me.

On a low level tour, I watched a group of primates at the edge of a forest and grabbed the controls for a closer look when I saw what appeared to be a humanoid stalking a sloth. The controls did not respond.

"Sorry, we are not on a search for the origin of man." Ralph took control and headed back into a polar orbit that allowed a fast overview of gross changes.

Slight climatic oscillation continued, but overall, Earth was cooling. We had reached the end of transfer of carbon dioxide from ocean and now the cooler ocean was taking carbon dioxide from the atmosphere. I checked the diagnostics and the methane levels had dropped, indicating to me that the ocean was no longer venting large amounts of methane. The land was well covered with plant life, and this mature state would not be capable of the rapid uptake of carbon dioxide from the atmosphere as it had in the past. My guess is that we were in for a long, slow cooling to Earth as we now know it.

In a close look at the carbon dioxide levels, I noticed an annual cycle of oscillation and looked back in history to see how it changed with time. It was extremely pronounced when the Arctic reached its tropical temperatures, then leveled out. I wished I could see the modern graph of carbon dioxide at Mauna Loa in Hawaii. Accurate data had been collected there for decades. My wish was granted and I studied it on the display.

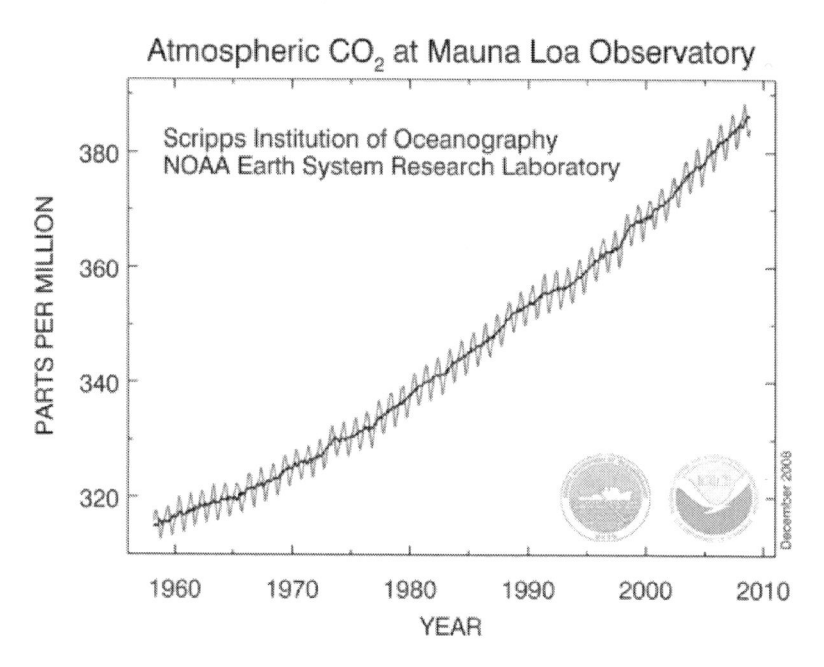

Atmospheric CO_2 at Mauna Loa Observatory
Scripps Institution of Oceanography
NOAA Earth System Research Laboratory

In my modern world, carbon dioxide levels were rising steadily, but the annual cycle was several times the annual increase. The annual cycles throughout ancient history were often undetectable, and when Earth was near its temperature maximum they were much larger than my modern times. Why? I had to think about that.

Ralph seemed impatient with my diversion and explained that the modern annual cycle was created by the excessive amount of land area and vegetation in the northern hemisphere compared to the southern. When growth started in the northern hemisphere's spring, carbon dioxide was consumed faster than all of Earth's respiration, decay, and energy processes could supply carbon dioxide and atmospheric levels fell. In the northern fall, total global growth was diminished and carbon dioxide levels increased.

"So, when the continents were evenly distributed between hemispheres, the annual cycle was negligible," I added.

"And when there was a tremendous polar area growth on one end of Earth with 24 hour per day sunlight, and little growth on the other, there were horrendous cycles."

A thought entered my mind and I chuckled.

"What's funny about that?"

"I just remembered a conservative radio newscaster announcing that 'carbon dioxide levels had suddenly dropped and scientists were perplexed' and then went into his energy company commercial."

"And we wonder why science is so difficult to understand," Ralph said in a remorseful tone and accelerated into time.

We went north and broke through a half a million years ago. I asked to return to the Canadian High Arctic to see the area where thick forests of Swamp Cyprus and Redwood had stood some forty million years before.

The islands south of the Arctic Ocean seemed to be in the same place and about the same size, but they only supported a spotty cover of grasses, tundra and small bushes. The lush forest had been burned or flattened and buried in rock rubble. Coal seams were visible on some of the steep hillsides. What a difference a few million years can make.

"Here comes the change you were waiting for," Ralph said.
It was late November and for the first time since the Earth thawed from its last snowball state, it began snowing. We went up to altitude and watched progress through time from seventy degrees north. Greenland was spottily covered with snow the first winter and within a decade year-round ice and glaciers covered its highest altitudes.

What a fast start toward glaciation, I thought. I wondered what had made the Earth cool so quickly.

"It was no one great thing. Equatorial desertification and the drifting of sand exposed an iron rich area that blew into the ocean to kick off a rapid biological growth and uptake of carbon dioxide at the same time the sun was in a cool cycle."

"Now the sun's energy is being reflected off of the snow and ice surface to further reduce heat input. We're in for a real freeze." The Antarctic must be leading the process, I thought. It's at a higher altitude and much colder. We were there in a heartbeat.

The continent was covered in snow and sea ice had filled in all its bays. Glaciers streamed from its mountaintops. It was the most spectacular glacier scenery I had ever seen. I was tempted to use my camera, but I wanted to save the one shot for the burning oceans.

Ralph chuckled. "Sorry, cameras are not allowed. Glad you came here?"

"It's awesome, thanks."

We went back north and watched the ice cap creep down from the Canadian north and glaciers appear on mountain ranges.

"What's different now, compared to the times that the Earth turned into a snowball?"

Good question. I was glad for the side trips. Revisiting the layers of carbonate rocks at Grand Canyon, seeing the coal seams in the Arctic and knowing that living matter had been turned into oil and gas made me realize that a vast amount of carbon had been stored below the ocean and land surface.

"There are a wide variety of carbon sinks. Last time carbon was stored only as soluble bicarbonates, solid carbonates, a lot of compressed vegetative seams, gas and oil under the oceans, and methane hydrate in the ocean floor." There were more differences. "The sun is warmer now and plant and animal life still flourish away from the snow covered areas."

"So what's going to happen?" He sped up the time as if to challenge me to guess before it took place.

"With the warmer sun, we will ice over more slowly. The polar oceans have long been cold enough to capture methane created from decaying vegetation and methane leaking up from oil layers as methane hydrate. As the oceans cool, less and less methane will be released to oxidize in the atmosphere. This energy loss will help to lower temperature. As the glaciers form, the ocean levels will drop. When the ocean levels reach the point where methane hydrate is released, there will be much less methane released than the last time. The temperature will rebound, but not enough to warm the ocean to the point that causes calcium carbonate to precipitate from dissolved calcium bicarbonate to release carbon dioxide."

He ran us at just about the speed I was talking. When the ocean levels reached their minimum we were over the Arctic Ocean.

"Most of the methane for this rebound comes from the Arctic. This area was the first to capture methane after we began cooling down and its relatively shallow basin is loaded with it."

The near shore areas looked like they were boiling. Once the process started, it spread throughout the area as the methane hydrate gave up its methane and it raced to the surface. Ralph assured me that similar action was taking place on a lesser scale all around the world, but this time the Arctic was the major supplier and there would not be any grand explosions.

I remembered my camera, and the goal to get a picture of the burning Arctic Ocean to convince Joe, Jeff, and Dale back in Cancun. I figured that Ralph and I had developed a tight relationship and he would let me get by with one photo. "Is there fire in the Arctic this time? Like the fires I saw on the test drive?"

We went up over the pole and spotted a fire near the Bering Strait. Ralph took me over for a good view and I stood to take the shot.

"No!" he shouted as the shutter snapped. Without a sound the windows blackened, and the cabin darkened, illuminated by only the blinking system monitoring lights and digital readouts. I could barely make out Ralph's somber face in the dim light.

"Trip's over," he flatly stated. He didn't sound angry, just disappointed. He shrugged his shoulders as though my action was inevitable and expected.

Needless to say, I was disappointed too. I really wanted to see the next ice ages. I clearly violated the rules, but it was just one little picture. If he knew everything, he must have known I had the camera. Why didn't he warn me?

Ralph turned to me with a very serious look. "There are limits to what I can do and there are freedoms that you possess that I cannot override. I cannot open the windows that you closed by violating the rules. You knew the rules. It was your choice."

"I feel like Adam holding an apple with a bite missing."

"Perfect analogy. When Adam, Earth's first man in the Judeo-Christian creation story, took a bite of the forbidden fruit from the tree of "the knowledge of good and evil," he became responsible for his actions. He had to work and was no longer cared for in the Garden of Eden. He was given superior intelligence, free choice and responsibility, including responsibility for the ongoing care of the planet. That creation story is a wonderful description of what man asked for and what the father of this universe wanted from man."

Wow! What was happening? Here I thought I'd spoiled the trip, and Ralph would be mad and unceremoniously dump me back in Cancun. Now what? Am I in a conversation with god? An angel of god? Who is this guy? Will I ever see the light of the day in 2010 again? He talks to the father of this universe? What's he doing with me?

The cabin lights came up to a fireside glow. Ralph turned and gave me a warm smile; his radiant eyes sparkled with kindness and care. "I'm not god. I am the Earth Keeper, a servant in the process of creating a peaceful, sustainable planet. The Earth team is dedicated to getting humans right. The Adam and Eve story, and other creation stories of the major and diverse religions, are a part of our doings, seeds we have planted to help humans enjoy and preserve creation's gift. Our goal is a human population that loves the universal creation, respects a diversity of worship of god, and cares for Earth."

Now this was really getting spooky. I had heard this message before. Somehow, I had just obtained and read three books by Thomas Berry, a recently deceased Catholic priest, student of global religion, and visionary of an environmentally sustainable Earth with a rich diversity of worship. This reading was out of my normal field and a little difficult to grasp, but I was moved by the connection between his spirituality and my science as it related to the ever-changing Earth. I had purchased several copies of his *Evening Thoughts* to share with friends until it went out of print and the price skyrocketed.

"I thought you'd relate to Thomas Berry's views, that's why I sent them to you."

So that's how he worked! I got some of them through Amazon, when I expanded an order to qualify for free shipping. Others seemed to pop off bookstore shelves at me. Was that the way he worked to influence me? How did he find me? Most important ... really, who is he? How does he fit into the total scheme? What is the total scheme? He had mentioned universes.

"You are not going to relax until you put me onto an organization chart. First you need some idea of where this universe fits in space and time."

I had thought that this universe was all there was and that it started from a big bang and was expanding. Ralph explained how this universe had cycled many times through an expansion and contraction. Not only that, but there were other universes within ours, or separate from ours, that expanded and contracted on differing cycles. All universes spawned planets with life. The goal of this whole exercise was the creation of a long-term planetary stability with humans living in harmony. Earth was one of the best examples Ralph had ever seen.

"I'd never imaged that there had been previous universes. This is mind-boggling."

Ralph went on to explain how he fit into the process. At the top was the mother of all universes. Each universe had a father of its universe and under the father were planetary teams for each planet capable of supporting a diversity of life. The planetary teams consisted of a designer, an executor, and maintainer. "These functions have decreasing power and I am at the bottom, the maintainer ... your tour guide and Earth Keeper," he concluded.

"Tell me more." I was so excited I could not contain myself. "Did you pick Earth? Was it assigned to you? Are all the team still in communication with Earth?"

Ralph's eyes shone brighter than ever when he described his team's search for a potential planet. Apparently they came into this solar system at about the time that we had started our trip and recognized the potential of a planet about the right distance from its energy supplying star, large enough to hold an atmosphere, and capable of maintaining a water based life system. His designer was especially excited about the presence of Mars. It would cool quicker and could be used to produce life well before Earth cooled to the point of condensing water. It could give them a jump-start on the competition.

The designer's power was great at the beginning, but limited. He had to make the tough early choices that the executor could implement. Once implemented, the executor's role was limited and Ralph took over. He observed Earth and humanity's progress, reported back to the team, and made minor adjustments. The team had apparently used most of its "father of the universe" granted divine power of intervention. Earth and its inhabitants were moving toward being on their own.

How did I fit into this? Ralph had certainly picked me out for something. What was it? What was I doing here? Wherever 'here' was now that I had shut us off from the outside a half a million years from my time on Earth. I was certainly not a powerful scientist. I did not really understand a small fraction of what Ralph had showed me on the fast trip through time. What did I know or control? What could I do? I thought I'd ask. "I know that I did not find you on that Cancun beach, you found me. Why me?"

"Do you understand methane hydrate's role in the recovery of Earth from ice ages?"

"Yes, but that is pretty simple and straightforward"

"Have you been reading about Earth's history and its temperature?"

"For many years, all I could find." He knew that. Where was he going?

"Have you seen any reference to methane release from methane hydrate as a reason for recovery from ice ages?"

I hadn't. That's where he was going. Scientists had been all around the area and somehow never connected the dots between ocean level decreases and the decomposition of methane hydrate. It is so simple and was right under their noses. How was it missed?

Ralph stood and paced the cabin, shoulders sagging and dejected,

"That's what we wondered. We see many places where man's rate of progress has dangerously outstripped his ability to understand its implication, but none are as serious as triggering another partial extinction. I'm glad you caught onto the seriousness of a slight oceanic warming."

Ralph was winding down and I feared the trip was at its end. I wished I could hang onto him for a few minutes. I had a graph of atmospheric methane concentration in my folder. After a steady rise to double its concentration over a period of a hundred years, it was showing a leveling off from 2000-2006. How much could Ralph tell me about this?

Ralph gave me a warm smile as a graph came up on the display. "It's been a great trip," he said turning to the graph. "I'll grant you one last request before I drop you in Cancun. I thought that perhaps you'd like the latest data."

The graph was a real shocker; methane levels had again jumped up in the past two years. "NOAA scientists attribute the recent increase to a warmer Arctic and wetter tropics." Ralph explained.

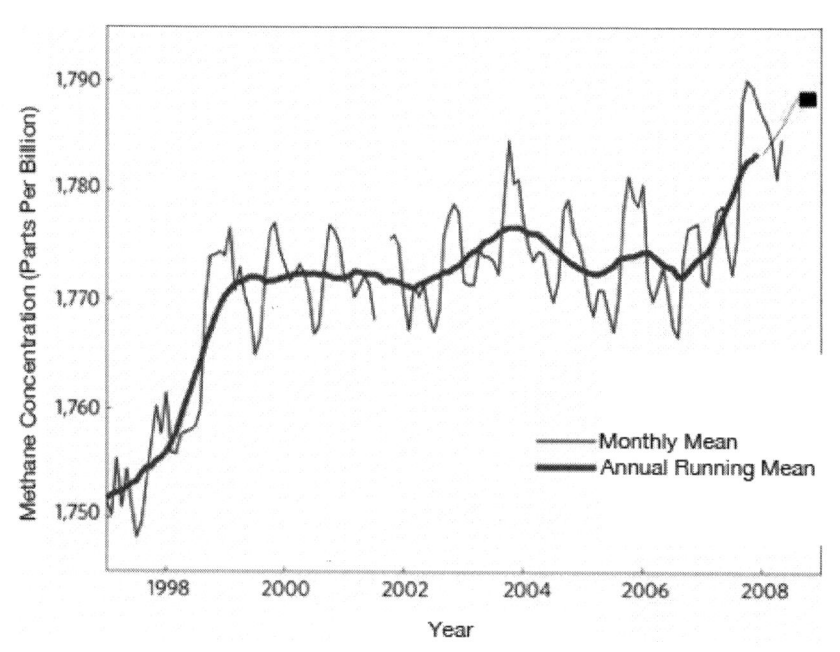

Ralph brought up another graph of methane atmospheric concentration over hundreds of millions of years. We were approaching historically high levels.

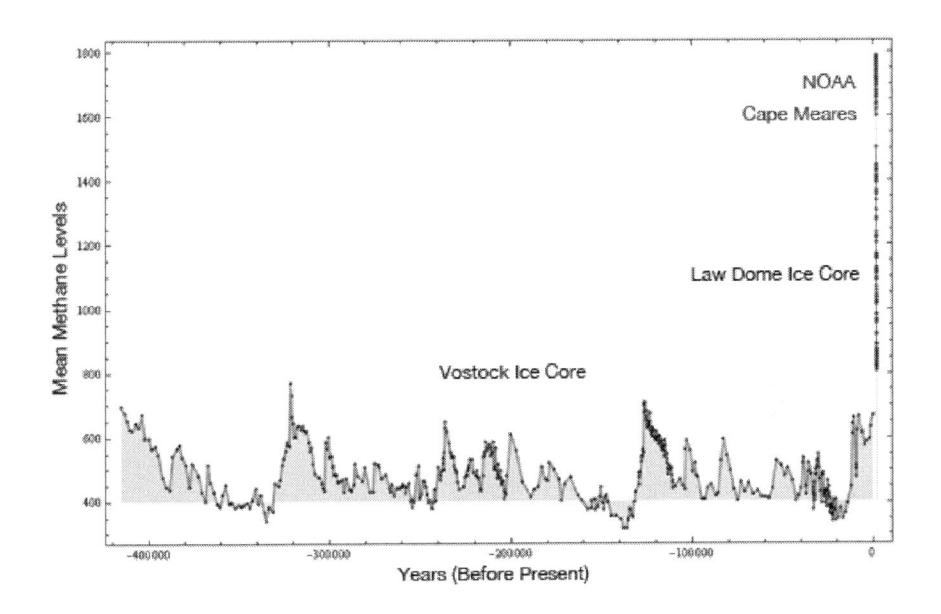

[Graph: Mean Methane Levels vs. Years (Before Present), ranging from -400,000 to 0. Data sources labeled: Vostock Ice Core, Law Dome Ice Core, and NOAA Cape Meares.]

"You are on the edge." Ralph looked at me with fear in his eyes. "I want you to survive in peace. That's my job! Please spread the word."
I held up my empty camera. "It would be a lot easier to convince people if I had a good shot of a burning ocean."

"Sorry about that. I'll do whatever I can to help?" He extended his hand and I knew the trip was over.

"A five minute clip of a burning ocean on national news would help." I smiled, started to say "thank you" and looked him in the eyes as I reached for his hand. His eyes flashed and I was gone.

I felt an elbow in my ribs. "Are you going to continue that global warming argument in your sleep?" Gloria asked. "Why don't you get up and go for your walk?"

Afterword:

It's me again. Wasn't that a glorious trip? I've often looked at modern photos of far away galaxies in wide-eyed awe and amazement, but nothing has compared to the thrill of spending a decade unraveling the mystery of Earth's recovery from an ice age. I had no idea that it would end up so simple and beautiful ... or that my path of inquiry would lead to a vivid understanding of Earth's miraculous transition from a collection of space dust to a beautiful and life-sustaining planet. I have reviewed the concept of decomposing methane hydrate as the major factor in Earth's recovery from an ice age with several experts who are in good agreement with this amateur's assessment. Before getting technical, let's review the process quickly just for enjoyment.

Earth accreted from space dust orbiting around the sun. Sister planets developed on either side of Earth. Mars, further out, developed an atmosphere, watery surface, perhaps life, and then froze. Venus, closer to the sun, developed a hot steamy atmosphere and will not cool to form liquid water. Earth's dense, oxygen free and carbon rich atmosphere cooled to create warm nutrient filled oceans. Oceanic life turned atmospheric carbon dioxide into organic material and oxygen. Oxygen converted dissolved oceanic metals into oxides, and then oceanic oxygen levels increased to supply oxygen to the atmosphere.

Atmospheric oxygen oxidized carbon monoxide and methane to carbon dioxide. The oceans knew no bounds for using up carbon dioxide and soon the atmosphere that had kept the planet warm was highly depleted of greenhouse gasses. Ice formed, polar oceans froze, and Earth's young continents glaciated. This should have been the end of the line as Earth progressed into a fate similar to that of Mars. But ... Earth had a better idea.

Organic matter that oceanic life created from carbon dioxide and water was being buried under sediments on the ocean floor. Over millions of years these sediment layers thickened. The lower layers, closer to Earth's hot mantle than to its cool oceans, were exposed to high pressure and temperature. Organic matter was converted to oil and methane. Methane, a gas, bubbled up through the sediment layers until it found a tight confining layer or cold water. If it found a confining area, a pocket of "natural gas" was produced. If methane continued to rise and found cold water in ocean floor sediments at depths greater than 600 feet of water pressure, it formed solid methane hydrate. This methane hydrate would save Earth from the freezer.

When the continents glaciated and ocean levels dropped, pressure on methane hydrate decreased and it decomposed into methane and water. The initially released methane dissolved in the ocean waters and was oxidized through biological processes. As ocean levels dropped and the pressure continued to decrease, more and more methane hydrate decomposed. Eventually methane bubbled into the atmosphere to begin a turn around of temperature, warming of the ocean, and release of more methane and carbon dioxide to recover from a glaciated condition. Methane hydrate was, and forever will be, the miracle compound that prevents Earth from permanently freezing.

Earth glaciated multiple times in its early history, joining volcanoes, wind and water in the process of turning Earth's crust into soils. Earth escaped its ancient freezing and thawing cycles by overshooting an ice age recovery to over-warm its oceans and initiate massive releases of methane from methane hydrate. In another miracle, by this time Earth had created sufficient atmospheric oxygen to support terrestrial life. As methane was released from the oceans, it was welcomed by a rapidly evolving array of plant and animal life that was eventually buried as peat and coal.

When Earth heated up to its maximum, about 55,000,000 years ago, the Arctic was covered in tropical splendor. In yet another miracle, the oceans ran out of carbon supply before the Arctic was cooked. Earth removed greenhouse gasses from the atmosphere and cooled. Before getting into the recent ice ages, I like to touch on spirituality.

Doesn't there seem to be a lot of miraculous happenings here? Earth was placed in the "just right" position to cool down to a "life-enhancing" optimal temperature. It has a large amount of water ... a thermal flywheel that dampens effects of temperature change induced by a variant sun, asteroid strikes, volcanoes, or underwater landslides releasing millions of tons of methane. Earth's orbital and tilt variations, along with variations in the energy output of the sun, invited periodic refreshing and renewing glaciations. Carbon was sequestered under lands and oceans to allow a moderate Earth temperature. Natural processes created a compound of methane and water at the ocean's floor to break a freeze up when necessary.

Do you wonder about these miracles? I grew up as a conservative Christian who believed that the Christian god created Earth in six days. It was only after I retired that I started thinking seriously about creation, spirituality, and religion. In my research for *Cold, Clear, and Deadly*, I was exposed to the spirituality of the Inuit. They, as most aboriginal tribes, possessed an ancient and rich relationship to a protective spiritual presence that guided their lives and relationship to nature. It was their handbook for thousands of years of survival in one of Earth's most hostile environments. Why did Christian missionaries take that away from them? Why do major religions claim a monopoly on god? Though my thoughts about Earth's development were engineering in nature, I often felt an overwhelming spiritual presence that is greater than any one religion's concept of god. I strongly recommend reading Thomas Berry's *Evening Thoughts*, or any of his essays on spirituality and the environment. This Catholic priest and student of global religions celebrates diversity of worship and before his death worked to get all religions to work together to support the sustaining of human life on Earth.

Earth may have been set in place by a divine order, but it follows the laws of chemistry, physics, and thermodynamics provided by that order. If we understand Earth's history, perhaps we can better predict its ... and our ... future. The recent ice ages have been well studied and are a good starting point for that understanding. The Vostock core of Antarctic ice has been analyzed back to 450,000 years ago.

This graph, in addition to temperature and carbon dioxide, contains methane concentration. Atmospheric methane varied between 300 and 700 parts per billion over the time span ... until man's industrial age. Let's take a closer look at atmospheric gasses, starting with the current graph of carbon dioxide with time.

Atmospheric CO_2 at Mauna Loa Observatory

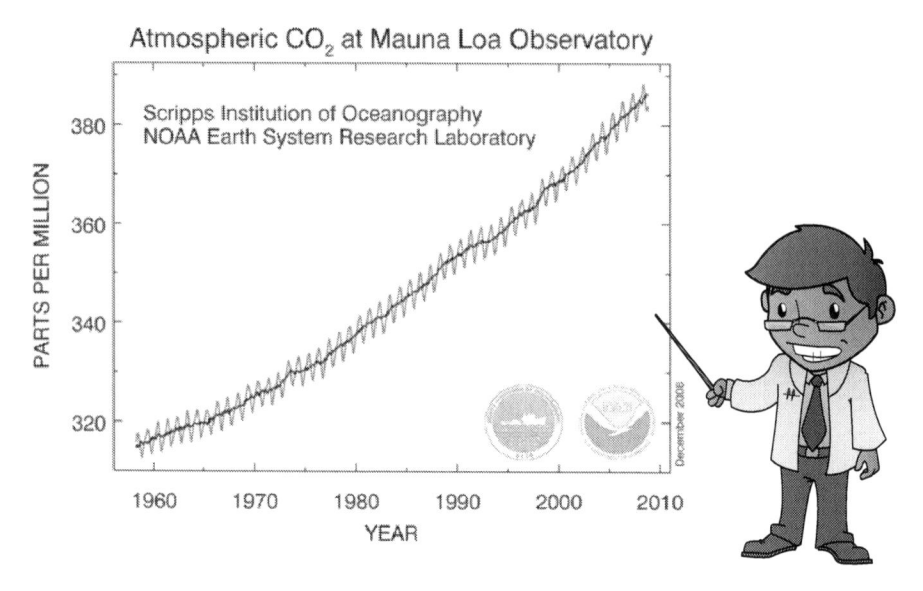

The concentration of a gas in the atmosphere can be compared to the level of water in a leaky barrel you are attempting to fill with water from a hose. If the leaks are large and the hose is small, the barrel will be nearly empty. If the leaks are small and the hose large, the level will be very high. Likewise, if the rate of addition of a gas to the atmosphere is increased, its concentration will increase. We have increased the addition rate of carbon dioxide to the atmosphere and not compensated with an increased removal rate, so carbon dioxide is generally increasing. In springtime, plant growth in the northern hemisphere removes carbon dioxide at a tremendous rate. This causes the atmospheric carbon dioxide to fall for a few months and result in our oscillating but steadily increasing concentration.

Now examine the Vostock methane graph. An increase of atmospheric methane concentration means that methane is being added to the atmosphere faster than it is being removed. At the start of every recovery from an ice age, methane concentration rises rapidly. Remember that any methane oxidized in the ocean, or catching fire as it is released from the ocean, will be released as carbon dioxide, not methane. Also, the higher the methane concentration in the atmosphere, the higher the rate of its oxidation to carbon dioxide will be.

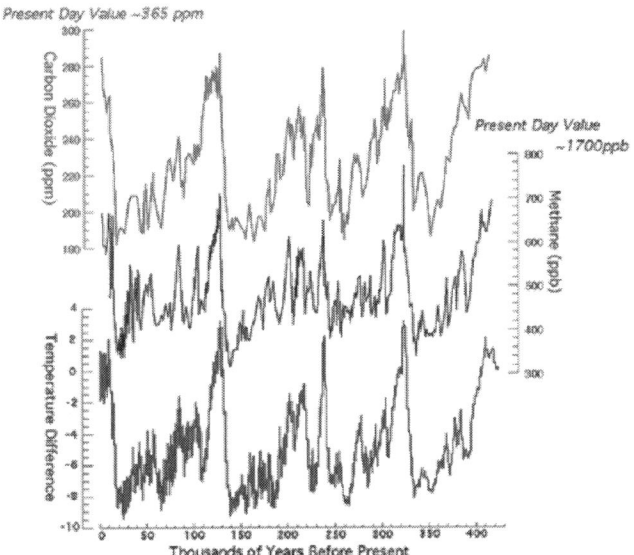

In all but the last recovery from ice ages, methane concentration rapidly decreased after reaching its maximum. As glaciers melted and water levels increased, methane releases from oceanic methane hydrate diminished and Earth cooled down.

The recovery of Earth from the last ice age took about 8000 years. During this time, Earth's temperature went up 18 degrees Fahrenheit and enough ice was melted to raise the ocean level 300 feet. This was all accomplished with an atmosphere that contained lower concentrations of greenhouse gasses than our current concentrations. After the most recent ice age recovery, the atmospheric methane level did not drop off in its usual rapid manner. Methane release rate stayed at a level, and supported a carbon dioxide level, that maintained the relatively stable temperature we have enjoyed for the past 12,000 years. This is yet another miracle that has allowed man to flourish. In the recent past, methane atmospheric levels have risen dramatically, to double the levels of pre-industrial man. This may mean that methane release rates have increased, that the atmospheric chemistry to remove methane has changed to where it not as efficient as it was in the past because we have challenged it with other pollutants, or a combination of both. As ocean ice cover retreats, tundra warms, and ocean temperatures increase, methane release rates will increase to propagate a positive global warming feedback and further warming.

As I write this, we are finishing our annual Cancun vacation where I learned more about how the sun varies in output and the difference between weather and climate. Our daughter Lori and her family visited for a week and we again enjoyed our five-mile beach walks. The Mexican government barged in a tremendous amount of sand and the beach is beautiful. It was a very cool year, the coolest in the 24 years we have been going there. Normally, we experienced strong east winds; powerful and steady winds that brought in rolling surf, howled through door cracks, and nearly blew us over as we walked between buildings. This year, the winds were from the north and blew the cold from the U.S. East Coast down to us. Meanwhile, Michigan was experiencing record high temperatures and the snow-covered state we left was under a wildfire alert.

Most of my old friends in Cancun came from conservative industrial backgrounds similar to mine and think that by studying climate change I have turned into a flaming liberal. Cancun's cold weather was proof enough to them that global warming was a hoax. Unfortunately, they, and the politicians who experienced the snowy 2009 Washington D.C. winter, are comfortable thinking that weather where they are at the time is climate . . . and will react accordingly.

For the past two years Dale, a professional marketing consultant and expert in data gathering and interpretation, had challenged my global warming thoughts and promised to bring me enlightening material. How refreshing to have a friend who will disagree with your ideas and conclusions and argue with respect. Dale had read my draft manuscript a year ago. He believed that warming was taking place, but detailed analysis showed it to be mostly at the poles in the winter and carbon dioxide did not correlate to be the clear cause. Could we pool his data handling skills with my applied engineering experience and come to some understanding? With the information he brought and the recent books, *Storms of my Grandchildren* by James Hansen and *A World Without Ice* by Henry Pollack, we hammered away at our differences.

Dale pointed out that the sun's irradiance variability is a whopping three watts per square meter, a very significant number when the industrial age forcing from all greenhouse gasses is estimated to be in that same range. The difference is that solar irradiance is measured as sun shining at the equator. The sun is only shining on half the world at a time and it strikes the non-equatorial areas with much less intensity. So, nearly 3.0 watts per square meter variation in solar irradiation only has the effect of 0.3 watts per square meter of solar forcing, still a significant amount.

It was no surprise to find from Dale that scientists studying the sun see the sun as having much more influence than scientists who study ancient climate and computer models. Sun scientists from Duke University and the U.S. Army Research Office concluded that 69% of the increase in Earth's temperature could be sun related and that some solar forecasts predict the sun's cooling off over the next decades to counteract greenhouse gas warming. (*Is Climate Sensitive to Solar Variability? Physics Today March 2008*) Cosmologists think that stars such as our sun are slowly warming during this stage of their life, but this is a very slow process and it could have ups and downs on its way up. The graph below shows the most recent satellite data on the sun's irradiance. It is indeed showing lower output and the current low we are in is the lowest and longest on recent record.

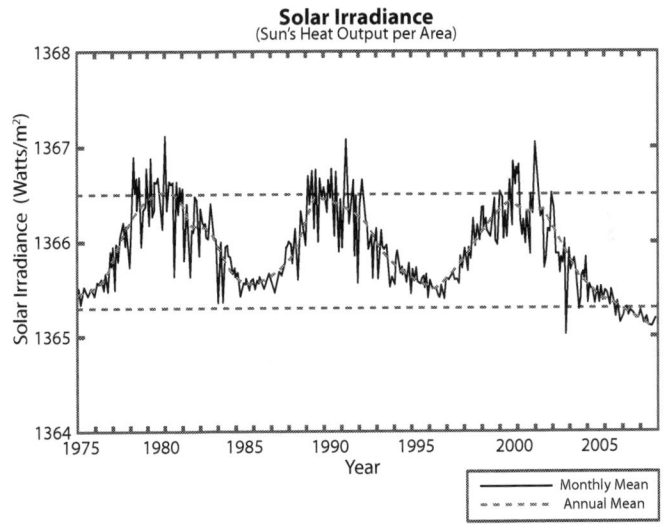

Solar Irradiance
(Sun's Heat Output per Area)

When Dale showed me the sun cycle in March of 2009, I did not appreciate its significance. I figured that by March 2010, the solar irradiance would be rising, but it has not. We are currently at the lowest and longest bottom of any recent eleven-year cycle. In the billions of years that the sun has been providing our heat, it has cycled in output and it would be a miracle beyond any described above if the sun does not turn around and increase its output over the next few years. It is very unlikely that the "hand of god" will reach out and dim the sun to save us Earthlings. We will have to do that for ourselves.

For the past 12,000 years Earth has been blessed with a rather stable temperature, optimal for man's development. An oceanic landslide releasing a billion tons or so of methane would be accommodated as Earth heated a bit and recovered. We are now pushing the warming throttle to the floorboard and any amount of methane added will only serve to accelerate warming. Methane is starting to seep from the tundra and Arctic waters. The ocean is warming and will cause decomposition of methane hydrate to initiate more warming and decomposition. In *A World Without Ice*, Henry Pollack documents the disappearance of global ice with time and states that the U.S. National Oceanographic Data Center is showing that since 1950, the oceans have been warming at a measurable rate. This effect is noticed at depths of up to 10,000 feet, but two thirds of the heat is taken up in the upper 2500 feet. The oceans are great thermal flywheels. If they are warming, climate is changing in a manner that may not be noticed by surface weather. There is no turning around of a warming ocean.

The only question is how fast Earth will warm. As the sun comes out of its cool cycle over the next few years, we will feel it. Many scientists cling to geological timeframes of hundreds of thousands of years, but never before in history has anything like our current condition existed. The overshoot of the last ancient ice age to produce the Cambrian warming happened naturally. We now have a similar condition, but man is transferring carbon from under the Earth and oceans and into the atmosphere to exacerbate it. Earth's capability of supporting 6.8+ billion of energy hungry humans will be seriously compromised in decades and generations, not centuries and millennia.

We will not destroy Earth. Earth will heat up until the carbon is removed from the sea to the atmosphere and most existing species are extinct. Then it will cool off, freeze and continue on with or without humans. Do we want to at least minimize the chaos for our progeny? Can we do anything?

Stewart Brand, in his *Whole Earth Discipline,* provides a very interesting assessment of what he deems necessary to turn around. Brand, an ecologist by training, was the creator and editor of "The Whole Earth Catalog," a highly successful environmental guide selling from 1968-85. He now heads a think tank providing long-term vision consulting to governments and corporations. He has evolved in his outlook and now views nuclear power as an absolute necessity. His book is very factual and thorough.

The amount of infrastructure needed to provide nuclear and other "green" power to satisfy the energy needs of nine billion humans and avoid going over the tipping point to runaway temperature would cover Australia. (And Brand's tipping point does not consider methane hydrate.) It is all in someone's back yard and will encounter opposition.

In his 2010 "State of the Union" speech, President Barack Obama stated the need for nuclear power, but so has every president since Jimmy Carter. These are "shovel ready" projects, but will they really happen? More likely, we will see "Cap and Trade," the mumbo jumbo that will send more carbon using jobs to coal-consuming third world countries. We will continue to import the cheap goods they supply and feel guilt free, as we will not be increasing our production of greenhouse gasses. In this wonderful economically flat world we have created, we have failed to see that it is environmentally warped. Globe travelling greenhouse gas pollutants are not lost when we outsource our toasters, furniture, electronics and steel; we give away the jobs and keep the pollution.

Global warming presents a real risk to our progeny. It is going to get hot, food is going to get short, and Earth is going to get ugly. This is a real risk. We must boldly take steps to non-carbon power, starting by following the French and Japanese example to create a global base load of nuclear power. It is the safest thing we can do to back yards of all continents as we learn how to directly harness the sun and the wind for a long-term power supply. Taking known manageable risks, so our progeny can deal with a lesser amount of chaos on a longer timeframe, is the least we can do for them. See http://www.coal2nuclear.com/ for a very enlightening approach to rapid conversion to safe nuclear power.

More difficult than the technical and economic problems of low carbon energy supply to the developed and developing world is the demand side. For the past two centuries, man has made great progress in science, governance, and religion. Globally, immense efforts have been made to feed the hungry, cure the sick, extend life, and embrace the oppressed. The result has been the domination of Earth by this single successful specie with every individual desiring to improve and lengthen his life through the use of use more energy ... energy we have continued to find and produce from sequestered carbon. We need to examine our "humanitarian" efforts and assure that they are used to improve quality of life and not quantity.

We have also shown amazing ability to use science to produce sufficient food to support this domination. Pesticides, fertilizers, genetic engineering, and irrigation have done wonders to improve yields, yet we have to continue to clear land for more agriculture and are forced to raise our food-producing animals in factory-like conditions. Increasing temperature will destroy usable agricultural land faster than our science can increase yields. Food production will diminish on land, as it already has from the oceans, and human population will have to decline. We will need new thoughts, plans, and cooperation to assure that diminishing food supply does not limit population through the agony of starvation.

My generation has gorged itself on the low hanging fruits of Earth's resources and man's ingenuity. We have created a condition that forces our progeny into massive actions requiring the cooperation of all global citizens in something other than global warfare or inaction that will result in chaos. This is new and risky territory. Let's hope and pray that we can find, encourage, and support the governmental, religious, and scientific leadership necessary to honestly assess and address our problems in a non-violent manner. We need the spirit of Thomas Berry to bring all our religions together, instead of driving us apart, as we work to create a world with a sustainable population powered with non-carbon energy.

Many books and websites provide sound advice for individuals to lower their carbon footprint and promote wind and solar energy. It is my hope and prayer that my efforts make it plain that we have to do more and do it quickly. I have offered several suggestions, especially rapid pursuit of nuclear power. If we really assess the risks, the warming risk is far worse than the risks surrounding nuclear power currently produced.

Another major arena in need of examination and improvement is the utilization of our wastes. We need to see our wastes as resources. The ongoing pollution of Chesapeake Bay by chicken manure that could be utilized as fertilizer is insane. Sending garbage long distances, like from Toronto to Michigan, to bury it when viable waste to energy options exist is wasteful. In days past, we sent our domestic animal, human, and industrial waste into oceans, lakes, and rivers. Now we spend trillions in infrastructure and energy to partially treat these wastes before throwing them into the same waters. This civilized engineering approach should seriously questioned. There must be millions of ways we can sustainably reduce and utilize waste.

In a justifiable reaction to past polluting practices we have focused on pollution elimination. Unfortunately, in a global economy this has resulted in transferring manufacturing to developing countries. Global warming pollutants are global in nature, so this is like throwing the baby out with the bathwater as we lose our jobs and retain the pollution. We need to focus on global sustainability ... the supplying of needs to the global population in a manner that retains the continuing ability to supply those needs.

As I was finishing this effort, I was going through the thrilling process of cataract surgery. Cataract surgery is now a painless five-minute procedure and within a half an hour I was seeing like I had not seen in decades. Man, in his compassion for his fellow man, has evolved to where he can make the blind to see, the deaf to hear, and the lame to walk. In fear and distrust of his fellow man, man could also quickly destroy his planet with existing nuclear devices. Just imagine what man could do if he dedicated a major portion of his creative energies and resources to a sustainable future.

I hope that we can learn how to question, collectively think, and act in a manner respecting the common good and global sustainability. Is our "post industrial" society really progress? Does exporting industry to avoid local pollution make global sense? The global economy has been good for the global economy, but is it good for the globe?

Thanks for listening and don't stop thinking and questioning. Then do as Lee Iacocca advised ... Lead, Follow, or Get out of the way!

Acknowledgements:

THERE IS NO WAY THAT I could have the luxury of research and writing time without the help of my wife Gloria who supplies all my needs, including making sure I am dressed before leaving the house. I have often said that I live in a comfortable cave with a high-speed internet connection and full maid, cook, and valet service. It's a great and much appreciated life.

A career of manufacturing chemicals and cleaning them up from the environment has supplied contacts with a multitude of industrial scientists and engineers, governmental regulators, and NGO activists with a variety of helpful expertise and opinion. Vacations in Cancun supplied regular contact with a number of friends over a period of years. Dale Hansson's critical and straightforward comments and insight have been especially valued.

Great Lakes environmental advocate, Dave Dempsey, reached across the industrialist/environmentalist gap to help me with *Cold, Clear, and Deadly*. Dave was the first to suggest that I apply similar thought to global warming.

My association with Michigan Technological University has been helpful in many ways. In 1995, I contacted Prof. William Rose and received helpful guidance. A few years ago, I was introduced to methane hydrate by novelist Clive Cussler who I owe a great deal of thanks. Once the methane hydrate cycling made sense, I went back to Michigan Tech and Dr. Rose. He directed me to Dr. Richard Honrath, an atmospheric scientist I had interfaced with on my PCB and pesticide global transport pursuits. Richard quickly grasped where I was coming from and I shared a copy of my draft manuscript. Unfortunately, Richard died in a kayak accident a few days later leaving nobody else with his interests and curiosity to take his place.

At the Byron Fellowship, a mentoring experience in faith and sustainability, I met Malcom Boyce, a retired global geological executive with a major oil company. Mr. Boyce spent his career searching for oil and was very familiar with methane hydrate. He reviewed two drafts, made significant contributions, and offered much encouragement.

Several other professors at Michigan Tech and several parishioners at the Milwood United Methodist Church have reviewed drafts. Bill Breyfogle supplied helpful comments and resource materials over a period of years. I have been fortunate to have our son Patrick, and Roger and Irene Booth, as vigilant draft readers. The comments of Rev. Wilbur Courter, and his friend, Dr. Carl Semmelroth, were much appreciated. Rev. Courter's understanding of a universal spiritual presence is something I wish was contagious and could start an epidemic.

Jim Ford's sharing of his draft history book and advice on publishing are also greatly appreciated. Jim led me to Fortitude Graphics where Sean and Sonya Hollins and Chad Sutton supplied an amazing amount of advice and talent to make a complex story more understandable.